ELECTROMAGNETIC SEPARATION OF RADIOACTIVE ISOTOPES

PROCEEDINGS OF THE INTERNATIONAL SYMPOSIUM
HELD IN
VIENNA, MAY 23—25, 1960

EDITED BY

M. J. HIGATSBERGER AND F. P. VIEHBÖCK
ÖSTERREICHISCHE STUDIENGESELLSCHAFT FÜR ATOMENERGIE GES. M. B. H., VIENNA

WITH 168 FIGURES

WIEN
SPRINGER-VERLAG
1961

ISBN 978-3-211-80575-6 ISBN 978-3-7091-5087-0 (eBook)
DOI 10.1007/978-3-7091-5087-0

Softcover reprint of the hardcover 1st edition 1961

Foreword

During the war years the techniques for electromagnetic separation of radioactive nuclides have made great progress. Ever since then further improvements have been accomplished to separate stable nuclides and in more recent years special attention has been given to the problem of separating radioactive nuclides, especially in connection with spallation and high energy fission studies. This poses slightly different problems, as the amounts of radioactive material may be very small indeed, highly diluted and the whole separation procedure has often to be a very rapid one, as some of the nuclides have a very short half-life.

On quite a number of the bigger machines very valuable operating experience has been gained and it was, therefore, a very good idea to hold a symposium covering all the aspects of electromagnetic separation, where technicians could discuss the recent advances in this wide field. It is to be welcomed that at this symposium all aspects were discussed, from the details of vacuum systems, as well as subjects like magnets, ion sources, scattering mechanisms, to the health physics problems connected, for example, with the electromagnetic separation of plutonium, as well as applications of mass separation.

These proceedings are, therefore, essential for all those who do work in this field or contemplate to enter into it.

Vienna, May 1961.

Dr. **Henry Seligman**
Deputy Director General
I. A. E. A.

Introduction

The art of separating radioactive isotopes by electromagnetic means has been improved considerably in the last decade. In order to achieve more accurate fundamental nuclear data and for a better understanding of nuclear fission and fusion processes, isotope samples of highest purity are required. The scale of work, so far, has a great range. Instruments are in use from the size of mass spectrometers or mass spectrographs up to large machines such as the calutrons. Thus, the sample weights amount from microgram to gram quantities.

Because of the activity of the separated isotopes and the initial feed material special precautions must be employed. Operational procedures involve remote handling and special health physics measures.

The separators are quite similar in the ion optical system but because of the above mentioned hazards, the technical solution of the machines vary quite distinctly from the stable isotope separators.

Only on two earlier occasions, Harwell 1955 and Amsterdam 1957, electromagnetic separation methods were discussed. At the Vienna symposium in May 1960, a number of European countries and the USA exchanged information on the latest experiences in the electromagnetic separation of radioactive isotopes. At this symposium, for the first time, the special problems were discussed which are connected with separating reactor or accelerator irradiated material. On the theoretical side, attention was given to isotope contamination factors.

It was felt that the publication of the proceedings of the conference could have a stimulating effect on scientists who did not have the opportunity to participate at the symposium.

The proofs of the papers have been checked by the authors who have also seen part of their discussions and comments. The editors acknowledge with thanks the cooperation and assistance of the authors and the Springer-Verlag in the preparation and presentation of this volume.

Vienna, Spring 1961.

M. J. Higatsberger and **F. P. Viehböck**

Contents

Chapter I

General Considerations Related to Problems of Electromagnetic Separations

Chapter II

Electromagnetic Separators Planned

Chapter III

Operating Experiences on E. M. Separators

Chapter IV

Progress in Ion Source Development

Chapter VII

Some Health Physics Problems Connected with E. M. Separations

Chapter VIII

Applications of E. M. Separated Radioactive Isotopes

Chapter I

General Considerations Related to Problems of Electromagnetic Separations

a) Separation of Irradiated Material

Electromagnetic Isotope Separation Before and After Activation in a Reactor or Cyclotron

By

Karl Ove Nielsen

Danish Atomic Energy Commission, Risö

With 1 Figure

In experimental nuclear physics, isotope-separated radioactive sources are often required. The possibility of employing the electromagnetic method for separation is very attractive, since this method, under ideal circumstances, will lead to an almost complete isotope separation in *one* step.

Two instruments often employed for activation are the reactor and the cyclotron, both of which can be used with great success in connection with electromagnetic isotope separation.

The types of electromagnetic isotope separators that are of interest can be roughly divided into two categories: large scale isotope separators (calutron), and laboratory isotope separators, such as the Scandinavian types.

It is well known that many isotope-separated radioactive sources have been produced by irradiation of samples of separated isotopes from calutrons, i.e. electromagnetic isotope separation before activation. If the laboratory isotope separator is available, it is often used for separation of radioactive isotopes, which means that the separation takes place after activation.

In the production of radioactive isotope-separated sources there are, in principle, the following alternatives: 1. laboratory isotope separator or calutron, 2. reactor or cyclotron, 3. separation before or after activation. These alternatives can be combined in several ways, but for practical reasons only a few of these combinations can normally be utilized.

Some general features of the various combinations will now be discussed.

In the case of separation after activation (separation of radioactive isotopes), the laboratory size isotope separator has been most prevalent. In this technique, the collection of the radioactive isotopes normally takes place by allowing the ion beam to impinge directly on a collector plate (direct deposition). Direct deposition results in negligible losses because the quantity of collected material is very small (no saturation effect), and the mass number of the collector material is normally low.

Collection by direct deposition is particularly advantageous in the preparation of sources for β-spectroscopy where it is desirable to have the activity on a very thin foil and concentrated on an area of 1/20–1/5 cm^2. Furthermore, the thickness of this radioactive layer must be uniform and preferably under 10 $\mu g/cm^2$. These requirements are normally fulfilled by the direct deposition method.

Activities Produced in a Reactor

If the activation takes place in a reactor, the (n, γ)-process is dominant. Fig. 1 gives the main feature of the two possibilities: Separation before activation, and separation after activation. In order to make a comparison, the conditions are so adapted as to obtain the same source strength (q curie) in both cases.

a) Specific activity

By separation after activation and collection by direct deposition there are only negligible amounts of neighbouring stable isotopes. In Fig. 1 is indicated the thickness of the layer in which the radioactive isotopes are deposited (r $\mu g/cm^2$). By separation before activation there will, practically always, be an accompanying amount of material consisting of stable isotopes. In many cases, the lower limit for such accompanying amounts of material will be $a_M \cdot q/S$ grammes — here S represents the specific activity attained by activation. This amount of material is much larger than that corresponding to the penetration depth r $\mu g/cm^2$ (Fig. 1).

b) The initial amount of activity

Since the efficiency ε of the isotope separator is approximately 1–10%, it is necessary to handle a larger amount of activity in separation after activation than in separation before activation (compare q/ε curie as against q/β curie); β is the efficiency of the source preparation method employed which, in many cases, is near 100%.

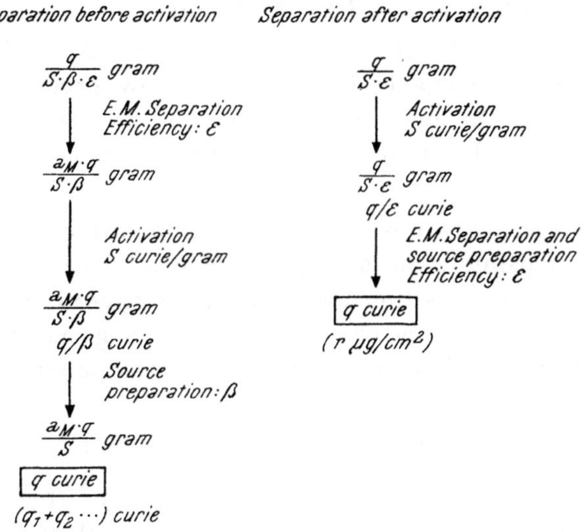

Fig. 1

c) Resolving power

The ion current corresponding to the radioactive isotope to be collected is lower in the case of separation of radioactive isotopes. This fact often gives a high resolving power and makes possible concentration on a small area, which is of great importance in cases where source preparation takes place simultaneously with collection.

d) Other activities and mass assignment

By activation, various kinds of reactions may occur which cause other radioactive isotopes to be produced than those intended. By separation after activation, apart from isomers, such varying activities are avoided; whereas, by separation before activation, they may often make an investigation difficult.

The combination of activation in a reactor and electromagnetic isotope separation before activation has often been used. However, it is obvious from the above that, for preparation of sources for nuclear spectroscopy, separation after activation is to be preferred in most cases. It will now be shown that it is only in a very few cases advantageous to undertake separation before activation.

Separation before activation

Preparation of isotope-separated radioactive sources by irradiation in a reactor of stable isotope samples from a calutron is a known technique and therefore will not be discussed.

Until now, a laboratory isotope separator has not been used in this manner, presumably because of its low production capacity. However, with the high neutron flux now available, it should be possible to use a laboratory separator for separation before activation in a reactor.

In this case, it is possible to use the following procedure: In a laboratory separator, the separated stable isotopes are collected by direct deposition on a thin foil which, after removal, is sent directly for irradiation in a reactor. Thus, an advantageous preparation of source is obtained simultaneously with the isotope separation.

Separation after activation

For separation after activation, it is usual to use the isotope separator of laboratory size, as previously stated. With the high neutron fluxes that are now available in reactors, there is, normally, no difficulty in obtaining suitable source strengths in spite of the loss in the isotope separator.

Provided that the loss by collection is negligible and that the ion current during separation is constant, the strength of the separated source Q_{as} is given by:

$$Q_{as} \approx 0.04 \cdot M \cdot J_i \cdot S_0 \cdot t \cdot e^{-\lambda t} \ \mu\text{C} \qquad (1)$$

where:

Q_{as} = the strength by separation after activation,
J_i = the total ion current in μA in the isotope separator,
S_0 = the specific activity in μC/μg of the reactor irradiated charge material at the time when separation is started,
t = duration of separation in hours,
T = half-life in hours,
M = Mass number.

Q_{as}, given by (1), has a maximum for $t = 1.44\,T$ (Thulin, 1955, and Nielsen, 1958), thus:

$$(Q_{as})_{\max} \approx 0.02 \cdot M \cdot J_i \cdot S_0 \cdot T \ \mu\text{C} \qquad (2)$$

For the isotope separator of laboratory size, the total ion current is of the order of $\mathcal{J}_i = 10^4 \cdot A_0 / \sqrt{M}$ μA, where A_0 is the area in cm^2 of the outlet opening of the ion source. In this case Q_{as} is:

$$(Q_{as})_{\max} = 200 \cdot A_0 \cdot \sqrt{M} \cdot S_0 \cdot T \ \mu C \tag{2a}$$

From eq. (2) *the attainable source strength is independent of the efficiency of the separator, whereas, the specific activity* S_0 *and the ion current* $\mathcal{J}_i$ *are decisive, and since only a small source area can be utilized, it is more the current density and not the total ion current which is important.*

The efficiency of the separator is determined for the amount of initial material to be used, and thus for the contamination as well.

For activation in a reactor, it is characteristic that the specific activity S_0 has an upper limit, whereas the amount of radioactive charge material, and thus also the activity, may be adapted according to demands.

The advantage of using separation after activation together with collection by direct deposition of the radioactive isotopes may be illustrated by a comparison between the strengths of two sources with the same area and thickness, one produced by separation after activation, and the other produced by irradiation in a reactor of isotope-separated material. The same reactor was used in both cases.

The source separated after activation has the activity Q_{as} as given in (2). Further, it is assumed that the area A of the source has been adjusted in such a way that the total collected activity may be used for the subsequent measurements.

Under normal operation conditions, $A \sim 2 \cdot A_0$, where A_0 is the outlet opening of the ion source. The ion source in a laboratory separator is often of such construction that it is easy to alter the area of the outlet opening, and thus it is possible to adapt A_0 to the actual purpose.

The thickness of the source is r μg/cm², where r is the penetration depth of the ions in the collector.

If S_0 is the specific activity attained by irradiation of unseparated isotopes in a reactor, the specific activity by irradiation of separated isotopes will be of the order of S_0 / a_M, where a_M is the relative abundance of the isotope. The strength of a source with the area A cm² and the thickness r μg/cm² is:

$$Q_{sa} = A \cdot r \cdot \frac{S_0}{a_M} \tag{3}$$

where Q_{sa} is the strength obtained by separation before activation in a reactor.

The ratio F between the two strengths of source is found from (2a), $A = 2 \cdot A_0$, and $\mathcal{J}_i = 10^4 \cdot A_0 / \sqrt{M}$ μA:

$$F_{\max} = \frac{(Q_{as})_{\max}}{Q_{sa}} \approx 100 \cdot a_M \cdot \frac{\sqrt{M}}{r} \cdot T \qquad (4)$$

Substituting

$$\sqrt{M} \approx 10, \quad T \approx 20 \text{ hours}, \quad \text{and} \quad r \approx 10 \ \mu\text{g/cm}^2,$$

then

$$F_{\max} = \frac{(Q_{as})_{\max}}{Q_{sa}} \sim 2000 \cdot a_M, \quad \text{for } t = 1.44 \cdot T \qquad (5)$$

If it is not possible to let separations continue until $t = 1.44 \cdot T$, it is found that:

$$F = \frac{Q_{as}}{Q_{sa}} \approx 200 \cdot a_M \cdot t, \quad \text{for } t < T \qquad (5\text{a})$$

From eq. (5) and (5a) it appears that sources produced by separation after activation are of a considerably higher quality than sources produced by irradiation of separated isotopes, except in the case of an isotope with an extremely small relative abundance (a_M). However, if a_M is very small, it will be advantageous, in many cases, to carry out separation before activation. Much can be gained by utilizing the possibility of extending the separation time until a suitable amount of material is obtained.

Activities Produced by Cyclotrons

In many cases it is necessary to use cyclotron bombardment in order to obtain the desired radioactive isotope. The conditions are very different from those of activation in a reactor. By irradiation in a reactor, the specific activity S has an upper limit, whereas the amount of activity can be chosen within wide limits according to requirements. The activity that can be attained by cyclotron irradiation has an upper limit.

Separation after activation

The amount of material that comes from the cyclotron target is so small that it is necessary to add some stable isotopes of the same element as the radioactive isotopes in order to achieve stable running conditions of the ion source (NIELSEN, 1957).

The possibility of adapting the amount of activity to the half-life and properties of the isotope separator does not normally exist.

The necessity of adding stable isotopes to the evaporation furnace of the ion source prevents the separation of radioactive isotopes with a short half-life. This difficulty has been overcome in two different ways in Stockholm and Copenhagen, respectively. These methods will be discussed later in this Symposium (UHLER and ALVÄGER, 1960, and SIDENIUS and SKILBREID, 1960).

The activity of the separated source:

$$Q_{as} \approx Q_{\mathrm{cyc}} \cdot \eta_t \cdot e^{-\lambda t}\ \mu\mathrm{C} \tag{6}$$

where Q_{cyc} is the activity in μC produced in the cyclotron, and η_t is the total efficiency of isotope separation. It should be noticed that in this case η_t affects the source strength directly. As previously stated, this efficiency had no direct effect on the source strength in the separation of reactor-irradiated material.

Separation before activation

Cyclotron irradiation often gives such small amounts of activity that sources produced by a subsequent isotope separation will be weak. In such a case, it will often be an advantage to carry out separation before irradiation. This can be illustrated by a comparison between the strength of a source produced by separation after activation and the strength of a source produced by separation before activation. It can be shown that the ratio between the strengths of the two sources is given by:

$$F = \frac{Q_{as}}{Q_{sa}} \approx \frac{a_M \cdot \eta_t \cdot m_u}{m_s} \tag{7}$$

where m_u is the amount of unseparated material that can be utilized on a cyclotron target, which is of the order of 5000 μg, and m_s is the amount of separated material available for irradiation in the cyclotron.

When a calutron is employed for separation before activation it will, usually, not be difficult to achieve the quantity of material that can be utilized on a cyclotron target, i.e. $m_s \sim m_u$. The ratio is then:

$$F = \frac{Q_{as}}{Q_{sa}} \rightarrow a_M \cdot \eta_t \qquad m_s \sim m_u \tag{8}$$

Equation (8) shows that in regard to source strength it is an advantage to undertake separation before activation, especially for small values of a_M.

If a laboratory separator is employed for isotope separation before activation instead of a calutron, a small value for a_M will make it difficult to achieve the condition $m_s = m_u$. For an isotope separator of laboratory size with an ion current $J_i = 1000/\sqrt{M} \cdot \mu A$, m_s is given by:

$$m_s = 40 \cdot \sqrt{M} \cdot a_M \cdot t \ \mu g$$

which, substituted in (7), yields:

$$F = \frac{Q_{as}}{Q_{sa}} \approx \frac{\eta_t \cdot m_u}{40 \cdot \sqrt{M} \cdot t} \qquad \text{for } m_s < m_u \qquad (9)$$

Substituting

$$m_u = 5000 \ \mu g, \qquad \text{and} \qquad \sqrt{M} \approx 10$$

then

$$F = \frac{Q_{as}}{Q_{sa}} \approx 10 \cdot \frac{\eta_t}{t} \qquad \text{for } m_s < m_u \qquad (9a)$$

Equation (9) and (9a) show that if a laboratory isotope separator is employed it is advantageous to separate before activation, since Q_{sa} will increase with t as long as $m_s < m_u$.

It should be noted that in these equations only the source strengths were considered and not the quality, as in the previous comparison. Generally, sources produced from activation after isotope separation have the following disadvantages:

1. The mass assignment is not certain. By irradiation in the cyclotron both (d,p), (d,n) and (d,2n) processes are possible.
2. The chemical separation carried out in connection with the source preparation is not complete, and, therefore, interfering activities originating from impurities may occur.
3. The sources are not of the same high quality as sources produced by isotope separation after activation by the direct-deposition process.

Separation—Activation—Separation

In the production of separated sources a combination of instruments which has proved advantageous in a few cases is the utilization of both calutron and laboratory isotope separator together with activation in either a reactor or a cyclotron. From the above discussion the most logical production sequence is as follows:

1. Separation in a calutron, giving relatively large quantities of material.

2. Activation in either a reactor or a cyclotron.

3. Separation in a separator of medium size, giving a high degree of purity and, simultaneously, ideal preparation of the source.

References

NIELSEN, K. O., Proc. 1955 Harwell Isotope Conf., Chapter 9 (1956).
THULIN, S., Ark. Fys., **9**, 107 (1955).
NIELSEN, K. O., Thesis (1958).
NIELSEN, K. O., Nuclear Instruments I, 289 (1957).
UHLER, J., and T. ALVÄGER, This book, Chapter IV; p. 193
SIDENIUS, G., and O. SKILBREID, This book, Chapter VI; p. 243.

Electromagnetic Mass Separation of Reactor Irradiated Materials

By

P. Weinzierl and **M. Tisljar**

Österreichische Studiengesellschaft für Atomenergie Ges. m. b. H., Vienna

With 3 Figures

The mass separator designed by the S. G. A. E. is mainly planned to serve two purposes:

1. Separation of samples of moderate activity (<1 Curie) after irradiation in the reactor.

2. Separation of stable isotopes for the preparation of targets for nuclear reactions studies and cross section measurements.

It is intended to deal in this paper only with the first point and to say only a few words on our other plans.

Isotope separation of a sample before irradiation in the reactor reduces the multiplicity of activities produced and thus in many cases facilitates the investigation of detailed decay properties. A considerable gain in the specific activity of the sample is achieved only in the case of a low natural abundance of the primary nuclide. High specific activity, however, is a main requirement for β-spectroscopy, β-γ angular correlations and other nuclear investigations. As most activities produced in the reactor are due to (n, γ)-processes, chemical separation methods are not applicable and only isotope separation of the activated sample can further improve its quality. A definite mass-assignment of the studied activity is a further advantage of this method.

For our further considerations one type of data is of interest: the amount of material with natural isotopic composition which is to be irradiated in the reactor (at 10^{12} n/sec/cm^2 thermal neutron flux) in order to achieve a certain activity of a specific isotope produced in the sample. Usually the saturation activity of the produced isotope can be attained. If the half-life is very long the activity refers to one year of irradiation. An activity of 10^6 decays/sec was chosen as a typical value for nuclear spectroscopy investigations. Figure 1 shows the number of isotopes against the amount of irradiated material in grammes. As we are considering the mass separation of the activated samples, this histogram includes only activities with a half-life of more than ten minutes. The full line shows the number of

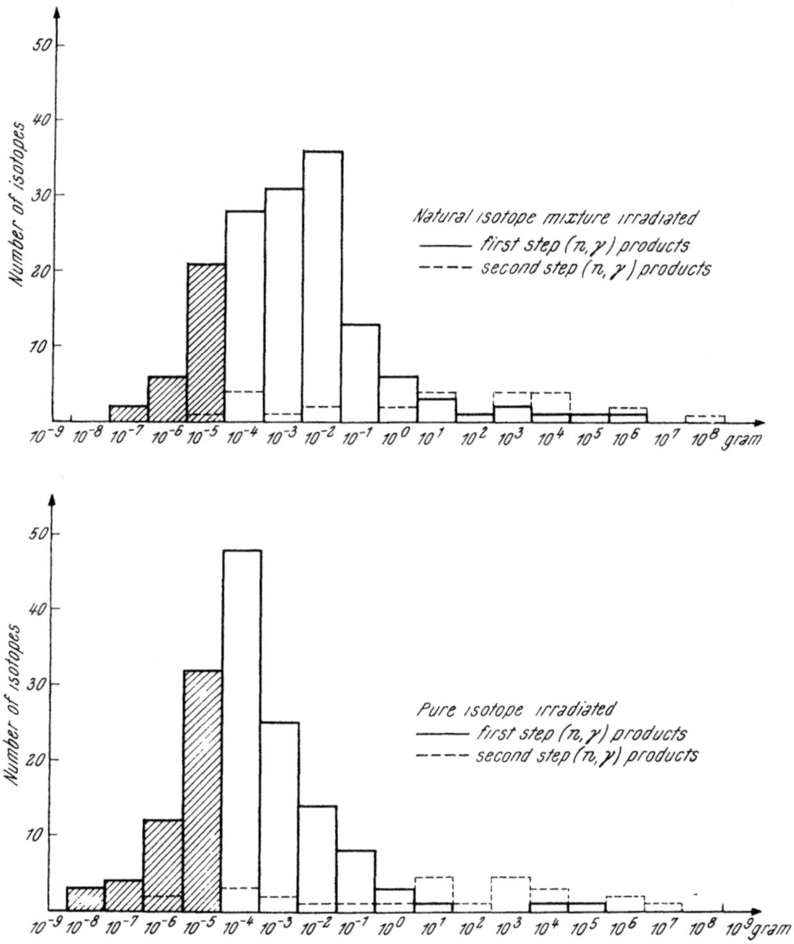

Fig. 1. Number of different isotopes produced by (n, γ) reactions as a function o weight needed for a certain activity of the sample

activities produced by a first step (n, γ)-reaction in the sample. The broken line refers to second step (n, γ)-produced activities. The lower part of figure 1 shows the same histogram for the case where, instead of a sample of natural isotopic mixture, a pure separated isotope is irradiated in the reactor. In the shaded region on the left-hand side of the diagram the specific activity is higher than 0.54 C/g and here good quality β-spectroscopy or similar investigations are possible without additional treatment of the irradiated material. The importance of mass separation after irradiation, even for isotopically pure samples in the reactor, can easily be recognized.

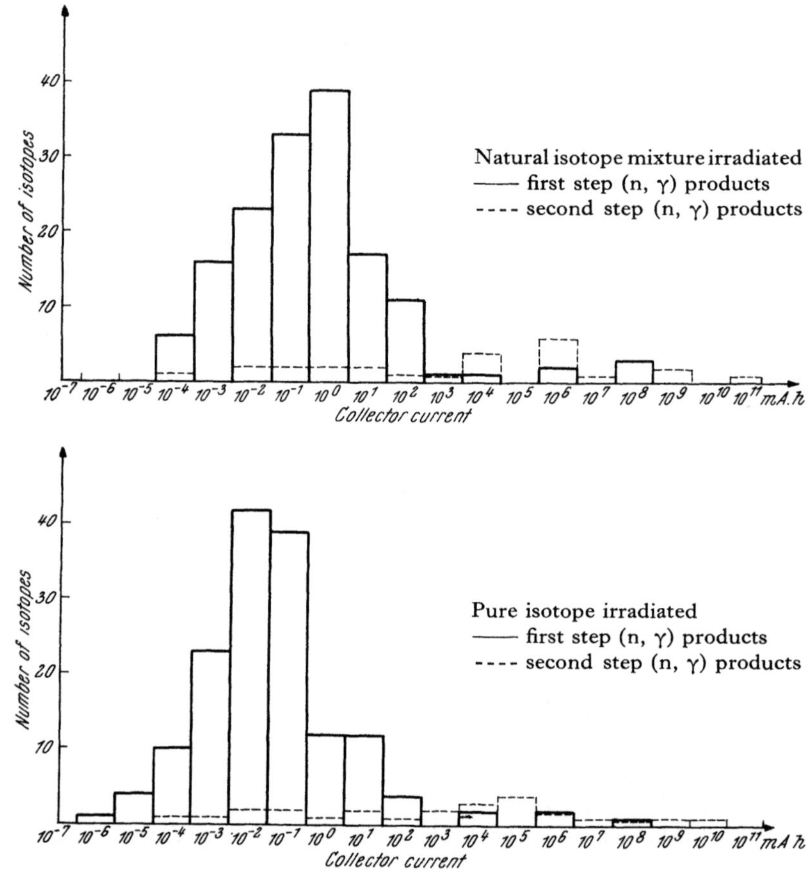

Fig. 2. Number of isotopes as a function of mA-hours of collector current needed for separation of a certain activity

Let us now assume for a moment that the half-life of the isotope to be separated is long as compared with the time of mass separation. Then the grammes quoted in our histogram are equal to the amount of material which has to pass our separator as ion current, if at the desired mass value an activity of 10^6 decays/sec is to be collected. The second figure shows the number of occuring types of activations [for a first and second step (n, γ)-process] with natural elements or pure isotopic material as charge for activation, but this time as function of mA · h of integrated ion current. In the next figure the decay of active material in the ion source as well as in the collector is taken into

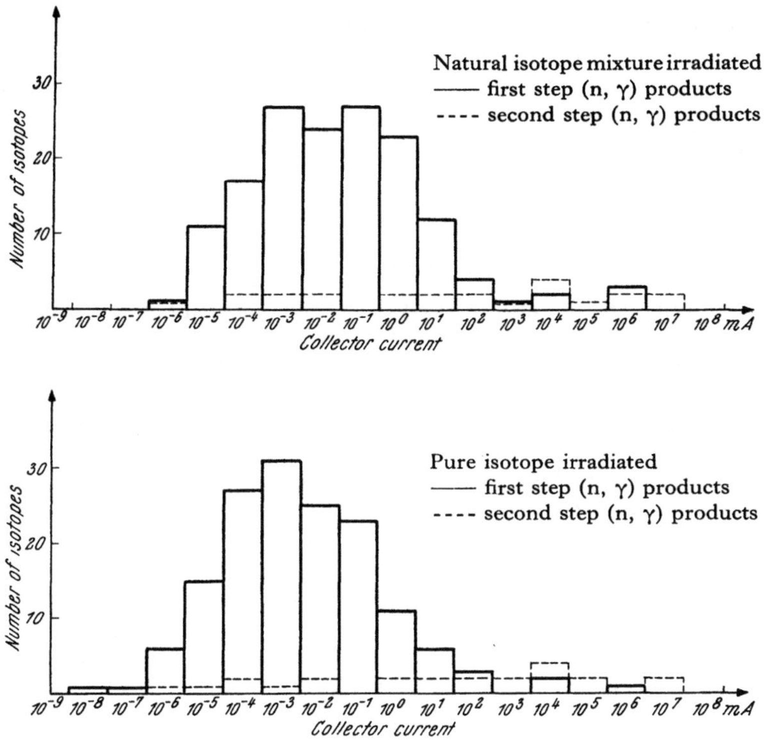

Fig. 3. Number of isotopes as a function of collector current for an optimum of collecting time

account. As shown by THULIN and NIELSEN the maximum activity of the sample is attained after a separation time of 1.44 half-lives of the separated isotope. The diagram shows the ion currents (in mA) needed to achieve the chosen activity after a separation time corresponding to the optimum value in each case. For long-lived isotopes, instead of this optimum value, a separation time of 100 hours was assumed. We think the usefulness of a mass separator for reactor-activated samples is clearly brought out by the diagram. For instance our planned mass separator with a maximum current capacity of 5 mA should be able to handle 86% of all cases occuring in this diagram.

May we conclude by a few remarks on other plans with the separator. The study of neutron cross sections of separated isotopes is of interest not only from the point of view of pure research but also for practical nuclear energy problems. With the improvement of fuel element technology and higher attainable burn-up the poisoning action of long-lived and stable fission products becomes important for fuel-cycle calculations. Therefore it is intended to use this mass separator also for the production of highly enriched samples of those isotopes which occur as stable end products of fission fragment decay chains.

Discussion

K. O. Nielsen: I am surprised that your calculations are based on a neutron flux in the reactor of 10^{12} n/sec/cm^2. I suppose that in many reactors you get 10^{14}. For instance, the neutron flux in the core of the DR 2 reactor at Risø is 10^{14} n/sec/cm^2 or a little higher. Besides, you talked about 5 mA ion current in the case of separating radioactive isotopes. I cannot see how you collect such high currents on a small area. Using a neutron flux of 10^{14} n/sec/cm^2 instead of one of 10^{12}, a factor of 100 is gained; then the required ion current is in the region of about 10 to 100 μA such as we have in the Scandinavian isotope separators.

P. Weinzierl: First point: I agree that higher fluxes than 10^{12} are easily attainable in a modern reactor core. But for short-time irradiations the regions outside the core are more practical. We plan to use our pneumatic system where we get less than 10^{13} n/sec/cm^2 at maximum power. On the other hand this pessimistic assumption is cancelled by the fact that our assumption on β-spectroscopy samples with 10^6 decays per second is rather low. With a short half-life one really needs the order of 1 mC to run the spectrum through. So I think it is a rather conservative choice of parameters.

Second point: We are aware of the fact that the pocket design must be such in order to trap the main ion current with the stable isotope. The radioactive and less abundant isotope will be collected on proper foils. There will of course exist difficulties caused by the big difference of ion currents on the collectors. This cannot be avoided however, when separating reactor irradiated materials. The current ratio is defined by nuclear data and the applied neutron flux.

K. O. Nielsen: I agree, radioactive isotopes can be collected on the foils, but I cannot see how it is possible to concentrate a beam carrying 5 mA on an area so small that collection by direct deposition can advantageously be used.

P. WEINZIERL: Let us assume we have a predominating isotope with mass 90 and then activate and obtain a product of mass 91. In our separator these two isotopes will be separated by about 20 mm at the collector. So I hope that it is possible to have here an exit slit for the main current near a collector foil for the active material.

K. O. NIELSEN: Experiences show that both in calutrons and in small laboratory isotope separators the current density is about 25 μA/mm². It means that with an ion current of the order of a few milliamperes, the area of the deposit region at the collector will be of the order of 50–100 mm². It is true that the current corresponding to the radioactive beam is weak, but it will be difficult to concentrate this beam without also concentrating beams of the corresponding stable isotopes.

P. WEINZIERL: I agree, but for cases where such an overall current is needed there will be no other way to collect the activity, and to remove it later and concentrate it by the usual methods of β-spectroscopy-source preparation.

O. SKILBREID: But then one will lose the great advantage of β-spectroscopy in connection with isotope separation. No other way of producing a sample for β-spectrometry is comparable with the direct deposition method. It gives the thinnest and purest sources.

P. WEINZIERL: I don't think that all good β-spectra are taken with isotope separators first.

O. SKILBREID: No, of course not, but the sources are much better.

P. WEINZIERL: Yes, I agree it is the ideal method. But if the extreme case occurs that one has a 5 mA beam of stable isotopes with the active sample 20 mm away there will probably be no other choice than to make some collections and to carry out chemical separation, finally evaporating this chemically separated activity in the usual way on a thin foil.

R. W. MCILROY: May I ask a question concerning these β-sources: what are the desirable properties of a β-source which one gets from direct deposition but not by an evaporation method? Secondly Dr. NIELSEN said that chemical separation introduces impurities. Does he find that separated isotopes have less impurities than ordinary laboratory reagents?

O. B. NIELSEN: The high quality of isotope-separated sources is partly due to the thinness and the uniformity. The efficiency of the chemical processing is about 50% but in general even lower. With 1 μg impurity coming from the air or water or reagents, there will be a tendency that this μg will crystallize in one or a few crystals. Such a μg in one crystal corresponds to an average source thickness of 10 mg/cm². This is not a very good source.

R. W. MCILROY: What thickness do you want?

O. B. NIELSEN: Well, that, of course, depends on the conditions. If we have high energy radiation then it is not so important but for low energy studies we want it below 10 μg/cm². I would like to mention another point that nuclear spectroscopists are quite happy to have pure activities much lower than 10^6 decays/sec. We have done quite satisfactory decay scheme work, where activities were of the order of a few 1000 decays/sec in β-spectroscopy. And especially when one deals with coincidence work one is not able to utilize the high activities because the counters run mad then, they only want a few μC corresponding to 10^5 decays/sec.

Note added in proof:

P. Weinzierl: With high resolution coincidence circuits there is no problem in allowing for activities $\sim 10^6$ decays/sec. In special cases work with activity as low as stated by O. B. Nielsen is certainly possible but the point here is to choose a typical value of desired activity.

b) Contamination and Enrichment

Theory of Isotopic Contamination in Electromagnetic Separators

By

Ch. Cassignol

Centre d'Etudes Nucléaires de Saclay

With 12 Figures

Introduction

At the Amsterdam Symposium on the Separation of Isotopes, three years ago, I described the two-stage version of the Saclay-Separator.

As expected it was found that the second stage improved the isotopic purity of the products of the separations and a tentative theory of isotopic contamination was proposed in order to explain it. This theory was based on scattering. Later it failed in accounting for other contamination manifestations and the basic physical phenomena were found to be known with insufficient accuracy. Systematic work was then carried out and brought to its end in November 1958.

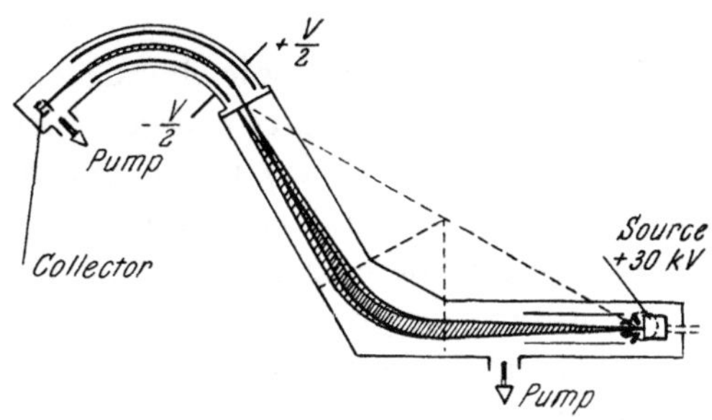

Fig. 1. Schematic View of the Saclay Separator

A detailed report has been written since the beginning of 1959. Two papers are now being submitted for publication in *Nuovo Cimento*. A short paper, by Dr. Bernas and myself, is also to be published in *Le Journal de Physique et le Radium*.

The present state of the work is a comprehensive and semi-quantitative picture of the isotopic contamination, which seems to be of a general application. When it is put into practice one can, on the one hand understand and justify the empirical operational rules of electromagnetic separation and improve the performances of existing machines; on the other hand one can establish, at least partly, the scientific principles for building up new separators, simple or complex. Therefore, it is of great interest to discuss the physical bases of isotopic contamination, before putting them into practice from the second point of view. One has to see if the present theory can be considered as right, either entirely, or partly, or not at all. This is one of the reasons why I submit it to your appreciation in detail. I am very interested in your comments.

I. Setting the Problem

The first point of emphatic importance is the notation point. The well-known enhancement coefficient has but little physical meaning

$$e = \frac{x'}{100 - x'} : \frac{x}{100 - x}$$

It is better to consider specific enhancement factors, defined by

$$e = \frac{a'}{b'} : \frac{a}{b}$$

a and b represent the isotopic abundancies of the main isotope A and a side isotope B in the source charge, a' and b' the respective abundancies in the collector pocket for A. When the processed element has several isotopes, several enhancement factors can be defined, each one is to be considered separately, not to be averaged in any way.

It is still more convenient not to consider the enhancement factors but their reciprocal quantities which I define as contamination factors. The advantage is that these factors, being much smaller than unity, can be treated as infinitesimal quantities; they consequently possess additivity properties, in respect of simultaneously acting distinct contamination mechanisms.

I am now going to define the contamination diagrams, the use of which will play an important part throughout the whole study, and to review with their help the empirical laws of isotopic contamination.

Fig. 2 represents the result of the isotopic analysis of a sample of Cd^{114}, collected in an early and bad separation. The masses are along the x-axis. Vertically the ratios a'/a, b'/b etc. of respective isotopic abundancies before and after separation. If we take as a unit, as will be done hence forth, the ordinate of point 114, the other ordinates represent the contamination factors of the respective side isotopes in the 114 one. We see from the figure that the closer the side isotope, the greater the contamination. The 116 contamination factor is higher than the 112 symmetrical one. We can determine a contamination function, which is practically Gaussian in the present case, at least for the lighter isotopes.

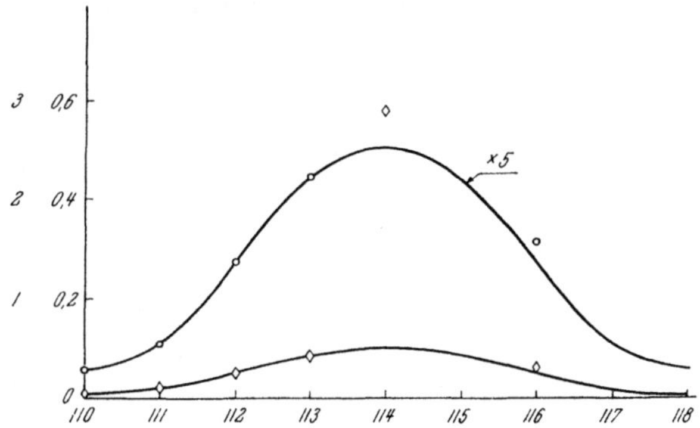

Fig. 2. Contamination diagram of Cd^{114}

Although the extrapolation is actually not safe at the 114 ordinate, one can see that unity cannot be the zero value of the contamination function. This remark allows us to define, if not to measure, the self-contamination factor, the physical meaning of which we shall see later.

The catalogues of enriched isotopes and a number of papers give similar results: influence of mass remoteness and preponderance of contamination by heavier isotopes. Nevertheless it appears that an opposite law is found for some instruments (in respect of the second criterion).

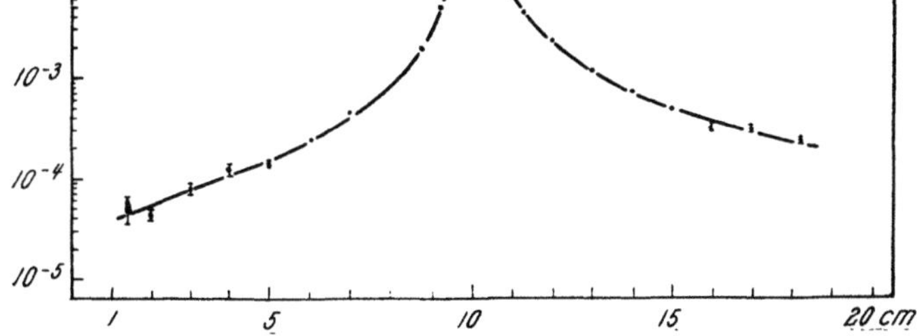

Fig. 3. Distribution of Kr^{85} on a metal plate (according to ALVÄGER and UHLER)

Fig. 3, taken from ALVÄGER's and UHLER's papers represents the distribution of radioactivity along a metal plate bombarded by a monoenergetic beam of radio-krypton 85 located at the collector place of the Separator of the Nobel Institute. Beside the main impact, the curve shows two tails the greater one towards the heavy masses; in this case contamination is higher for lighter isotopes.

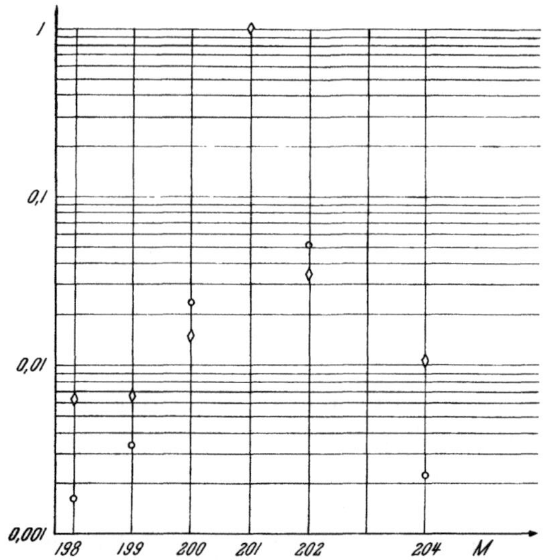

Fig. 4. Contamination diagrams of mercury 201

Fig. 4 represents two superimposed diagrams of mercury 201, obtained by the two-stage machine of Saclay, with distinctly different pressures inside the tank. One can see that the better value of the pressure for the remote isotopes 198 and 204 is worse for the closer neighbours 200 and 202, and vice versa.

These are the principal data to be accounted for and physically explained.

The figures emphasize the necessity to take interest in any information provided by any side isotope. It appears that the shape of the contamination diagrams can be of major interest and the whole study is built up accordingly.

Rather than consider the average enhancement or one specific contamination or enhancement factor, and adopting a simpler method than to consider all the contamination factors, I shall examine the comparisons of the contamination factors between themselves, the mean level of contamination being for the moment considered as a subsidiary matter.

If the contamination mechanisms have any constancy, contamination does not change except in level, and the contamination diagrams can be superimposed by the help of a vertical translation, if they are drawn in a semi-logarithmic diagram. If this is not the case, the shape of the diagrams is changed and contamination suffers a qualitative change.

The changes of the shape of the contamination diagrams we shall have to deal with are of two types, already present in the above figures. The first type concerns, for the same mass differential, the inequality relations between the contamination factors of heavier and lighter side isotopes. The second one concerns the range of the diagram, the more or less abrupt character of the contamination function.

As an illustration I anticipate the final result. We shall meet two principal types of isotopic contamination which are present simultaneously: scattering of the ions by the residual gas and instability of the ion sources. The first one is a long-range, smooth-variation contamination, higher for lighter than for heavier equidistant side isotopes. The second one is a short-range, abrupt-variation contamination higher for heavier than for lighter equidistant side isotopes.

II. Trivial Mechanisms and Energy Spread

Before considering these mechanisms, I must get rid of some mechanisms which actually contribute to the contamination, but can be annihilated by proper and specific care. Such are: vapour migration to the collector, *rebouncing* of ions on the walls of the chamber or inside a collector assembly, sparkover in the acceleration gap, poor stability of the supplies, misalignments, bad setting, mistakes of the operators, memory effects. I shall assume that they are, reduced to a negligible level, as they can be, and I shall no longer speak of them.

The energy spread of the ions is sometimes more or less implicitly taken as responsible for some isotopic contamination. In the recent book by Koch, Dawton, Smith and Walcher, we can read in the chapter dealing with the Saclay Separator (p. 88): "Obviously neutral atoms originating from the first stage will be sorted out by this procedure; but in addition all ions which do not have the proper energy will go to the walls of the chamber due to the energy dispersion of the electrostatic field . . . It is seen that the electrostatic analyser considerably improved the isotopic purity of the separated samples; nevertheless the author suggests that two magnetic stages in cascades would be still more favourable."

This last statement implies that energy spread is of little importance in respect of the other mechanisms, since two magnetic stages in cascade are not more efficient than a simple instrument as far as energy spread is concerned. It is consequently of emphatic importance to determine unambiguously the contribution of energy spread to isotopic contamination. The second electrostatic stage of the Saclay separator is an ideally suitable instrument for this purpose, provided it is satisfactorily operated.

Two sequences of experiments were devoted to this problem. The first one with a source running in a stable regime convenient for separation. The second one with the source running in a hashy regime. In both sequences, I could not detect any energy spread that would account for any isotopic contamination.

In the first sequence, the operated body was molybdenum, processed in a sputter arc source with argon as a support gas and a side-band. Let us assume that energy spread is responsible for some contamination. The relation of the first stage

$$MV = KB^2R^2$$

shows that the energy spectrum of the ions penetrating the second stage is discrete, as the mass spectrum is. Assuming that the resolution of the second stage is sufficient, we must record, when scanning the electrostatic voltage, a discrete spectrum, in both the mass and energy spectra. Actually when the electrostatic voltage is scanned, one records a single current peak, rapidly falling to zero on either side. Then the broadness at the foot can be safely estimated. It corresponds to a dispersion of 0.3% in energy. It is sufficiently small to ascertain that the discrete spectrum is not imbedded in the current peak. Moreover, this broadness can be entirely attributed to the instrumental width of the second stage, including the width of the entrance and exit slits, aperture defect, and residual misalignments. It follows that energy spread can in any case produce isotopic contamination but to a negligible amount compared with other mechanisms.

III. Scattering as a Mechanism of Isotopic Contamination

The work beams of the E.M. Separators are atomic ions of unitary charge beams, the energy of which is about some tens of keV. The theory of the collisions of ions and atoms in this energy range and thermal atoms and molecules is poorly elaborated. A few experimental data are nevertheless available and will be mentioned now.

When a collision occurs, the velocity and the energy of the fast particles are only negligibly modified. By far the most frequent type

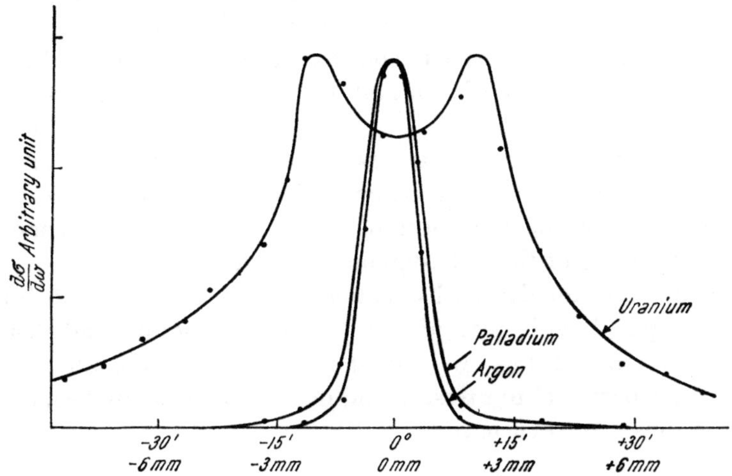

Fig. 5. Angular distribution of fast neutrals originating from scattering in Argon of unitary charge, 30 keV energy ions of A^{40}, Pd^{108}, U^{238}

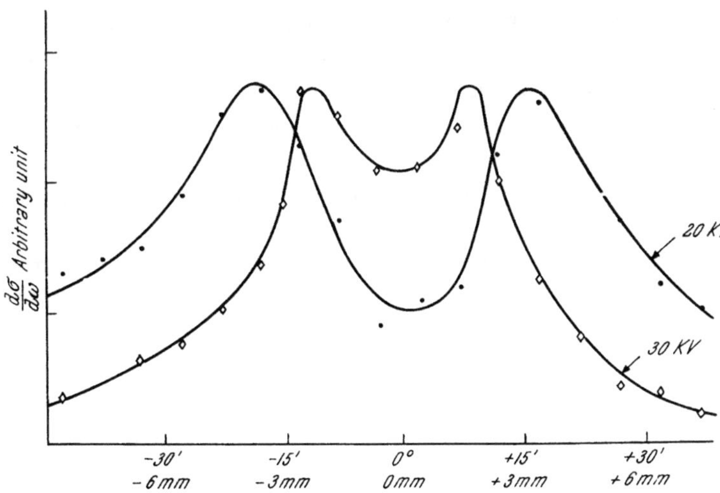

Fig. 6. Angular distribution of fast U^{238} atoms originating from scattering in Argon or unitary charge U ions, the energy of which being 30 and 20 keV

of collisions is charge exchange or neutralisation. But stripping collisions and collisions without charge change for the fast particles are observed. A few cross sections are tabulated but it is unnecessary to know them when using a comparison method.

The scattering angles are small. They are greater, according to FEDORENKO, for stripping than for neutralisation. They are greater for neutralisation of heavier constant energy fast ions than for lower energy fast ions of constant mass, the scattering gas being unchanged.

The scattering collisions can lead foreign ions or atoms into the collector pocket for a given isotope. They can also lead scattered ions or atoms of the same isotope. (One can then understand the physical meaning of the self contamination factor.)

Provided that the density by volume of the residual neutrals is high when compared with that of the ions of the beams, and that the absorption remains low, we can state that a given contamination factor is proportional to pressure and does not depend on the intensity of the beams. We then have:

$$c = \lambda p$$

But we also have

$$\lambda = \lambda_{(1,0)} + \lambda_{(1,1)} + \lambda_{(1,2)} \ldots$$

the first $\lambda_{(1,0)}$ corresponds to neutralisation contamination
$\lambda_{(1,1)}$ corresponds to contamination of scattering without charge change

The $\lambda_{(1,2)} \ldots$ correspond to the stripping types of contaminations. The greatest λ is the first one, i. e. the scattering contamination is mainly of the neutralisation type; so, if the mass of the ions is not too high, there is a sharp preponderance of contamination by lighter isotopes. As a matter of fact, we can then neglect the scattering together with neutralisation (it is often neglected in the literature), and it is obvious that the efficient regions in respect of this type of contamination are close to the exit from the magnetic field, and that the neutralized ions leave the unperturbed trajectories, and can contaminate only heavier isotopes. (Inversely, a given isotope can be contaminated only by lighter isotopes.)

This mechanism appears to be preponderant in the Scandinavian separators.

The stripped beams can be observed very easily. When operating the second stage at one half or one third of the ordinary voltage, one can measure a current which increases when the pressure is raised in the first stage and which corresponds to electron losses by ions in the second straight part of the trajectories (after crossing the magnetic field). I had already mentioned these observations at the Amsterdam Symposium.

IV. Hash Contamination

When the pumping speed of the vacuum supplies is made larger and larger, it is known for most types of apparatus that the resolution is destroyed and that the intensity to the collector diminishes. The beams burst and are lost. The pumping speed of the Saclay separator being too small we were able to observe this phenomenon with the source running but in a somewhat hashy regime.

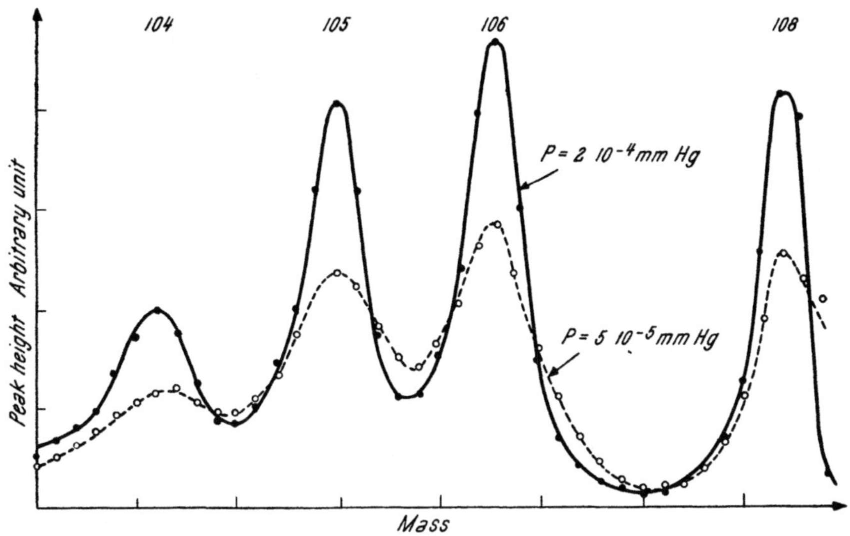

Fig. 7. Mass spectrum of palladium, showing the influence of the pressure in the separator chamber on resolution

It follows that scattering is not a trivial mechanism, and that the actual behaviour of the pressure is rather intricate.

Almost all E. M. separators work with a diode acceleration gun, the first electrode running at a negative potential, in order to keep the electrons within the beams, by preventing their being trapped by the ion source. The mechanism of space charge compensation pointed out by BERNAS works in this simple way when the beams are ideally stable, and we call it static neutralisation. It cannot be impaired by the lowering of the pressure in the main vacuum chamber.

The hash is a particular type of instability of the sources of E.M. Separators which run in a magnetic field. The hash of the discharge is transmitted more or less exactly to the beams as fluctuations in density and intensity, and results in a random space charge residue. Experimental work in the early time of the Manhattan project has shown that this residue can mostly be lowered by raising the main pressure, that its average value is positive, and that there is no experimental evidence for negative instantaneous values.

This can be easily understood. When the intensity is steady, the space charge is statically neutralized. When the intensity increases, an excess of positive space charge of fast ions appears, which is neutralized in the same way as in the static process, within a time of the order of 10^{-4} sec, for reasonable values of pressure, and which, as a first approximation, varies like the reciprocal pressure. The average value of the residue is in the same way proportional to the reciprocal pressure, and of course, proportional to the intensity of the beam.

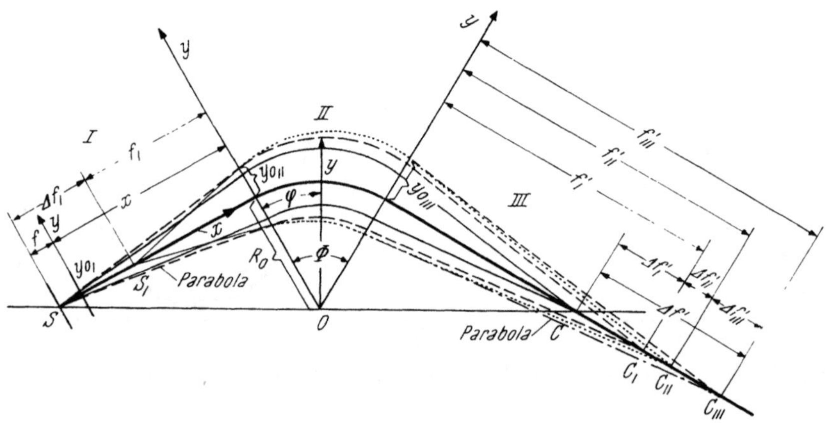

Fig. 8. Two-dimensional approach of the influence of the space charge on the focussing properties of an E. M. Separator (according to Becker and Walcher)

When the intensity decreases, there appears a negative space charge consisting principally of slow electrons, which vanishes by migration to the walls under the electrostatic action of the electrons themselves, and by the temperature. The time constant can be estimated. It does not depend on the pressure, and is considerably shorter than the former one.

It is smaller than the transit time of the ions across the machine, that is about 10^{-5} sec. It follows that the negative residue and the negative space charge potential excursion can be neglected. The space charge potential is positive or zero.

When the space charge is not perfectly neutralized, the residue being positive, the optical properties of the separator are altered (Fig. 8).

The space charge residue fluctuating between zero and a maximum value, the focus point fluctuates between the ends of the above line (C and C_{111}), with a stop at the ideal focus C which is the practical focus point, the entrance slit of the collector. The average loss in resolution involves some contamination, which is in principle symmetrical (equal for heavier and lighter isotopes).

A more intricate way of manifestation of the space charge will be suggested now, in order to explain, when superimposed on the former one, the preferential contamination by heavier isotopes. It consists in considering the slowing down of the ion beams when the space charge is not compensated, and when the space charge potential is not zero, but positive.

Fig. 9 represents in an ideal plane case the space charge potential created by a homogeneous positive space charge. When an ion, accelerated through V volts, travels inside a space charge potential of v volts, its kinetic energy is $V - v$ electron volts; this is lower than its value when the space charge is compensated, when it has its full accelerating value.

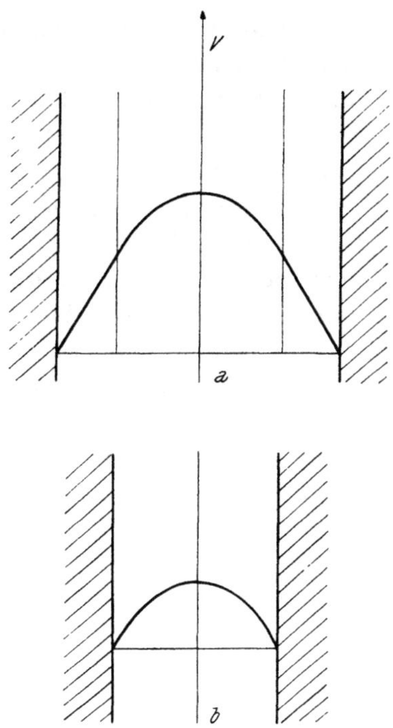

Fig. 9. Space charge potential diagram
a) the walls are remote from the space charge
b) the walls are close to the space charge

Along the straight paths of the ions this effect is of secondary importance. But inside the magnetic field, the slowing down entails a decrease of the curvature radius of the ion trajectories, that is a side displacement of the beams at the collector towards the light mass side. The displacement of the point focus is not as in the former figure along the mean beam, but oblique, and the feet of the beam are no longer symmetrical. The greater foot being on the light-mass side, the corresponding contamination is greater for heavier isotopes.

I have no direct proof that this mechanism comes in at a sufficient level to explain the preferential contamination by heavier isotopes. Nevertheless I do not know any other mechanism to explain it (remember that energy spread was formerly refuted) and its consequences are all verified by experience.

Moreover one experimental result can be considered as a quasi direct proof for the above mechanism. Actually it suggested it. It consists in considering the influence of the remoteness of the beams from the walls.

When the walls are brought nearer to the beam, the space charge potential excursion is lowered together with the asymmetrical part of contamination. A note added in proof in the paper of LIVINGSTON and MARTIN in the Amsterdam Symposium can be interpreted in this way. It concerns the improvement in enhancement of the L. I. I. S. in Oak Ridge for U^{235} in respect of 238. Two grounded electrodes, normal to the magnetic field, distant 11 cm from each other, closely surround the beams. Before these electrodes were used the beams travelled in the chamber the height of which was about $3\times$ larger. It can be stated that if my assumption is true, the above mentioned operation must on the other hand impair the enhancement of U^{238} in respect of 235. But LIVINGSTON does not give any information on the subject.

The loss in resolution due to the hash involves a typical isotopic contamination, the characteristic of which is: short range contamination increasing with the intensity and the reciprocal pressure. We shall write simply $C=\mu i/p$. For two equidistant isotopes, the contamination is greater by the heavier one.

V. Complete Scheme of the Isotopic Contamination

We now can try to see if all the known phenomena of isotopic contamination can be explained by condensing all the scattering and hash mechanisms into a single comprehensive framework.

The representative formula is $c=\lambda p+\mu\frac{i}{p}$.

The validity of this formula was controlled in a few test separations of Cd^{113}. This isotope is quasi symmetrically surrounded by reasonable abundance isotopes. The regime of the source being kept constant as much as possible, the chief parameter was the pressure inside the separator chamber.

In the first experiment, the pressure was very high (argon gas was injected into the chamber). The scattering was then very efficient and the hash inefficient. The characteristic of the former must predominate. The preferential contamination by lighter isotopes is evident. $[C_{112} > C_{114} - C_{110} > C_{116}]$

In the second experiment, the pressure was very low (no argon was injected). Then the hash predominates, and the preferential contamination by heavier isotopes is evident. $[C_{112} < C_{114} - C_{110} < C_{116}]$

In the third experiment, the pressure has an intermediate value and the respective characteristics equilibrate: the diagram is almost symmetrical.

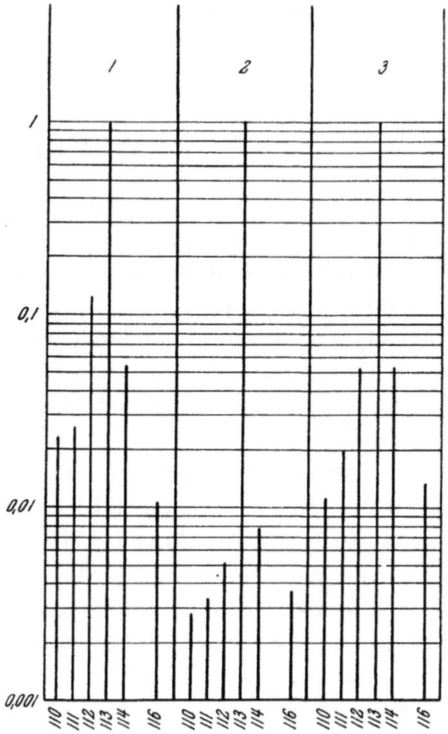

Fig. 10. Contamination diagrams of Cd^{113}

Thus the proof of the validity of the formula $\lambda p + \mu \frac{i}{p}$ is given.

I sum up the conclusions: leaving aside a few mechanisms which can be corrected by specific precautions, isotopic contamination is due to the simultaneous action of the scattering and of the hash of the sources. These mechanisms cannot be treated as trivial since the needs for correction in respect of pressure are contradictory.

The formula could, of course, be improved, but this does not appear very necessary. A quantitative formula cannot be derived, since the fundamental parameters of a separator are never well known. There are many gaps and much inaccuracy in the tabulation of scattering cross sections. The hash is not a measurable quantity. The pressure itself cannot be safely connected with the indication of a ion gauge for quantitative deductions. For semi quantitative deductions, the above expression is sufficient.

VI. Applications

I cannot treat the applications in detail. It would take too long. They involve principally:

1. Applications to simple machines: the practical operational rules for simple instruments: optimum pressure; compromise between intensity and resolution, geometrical rules for building new machines, comparisons between the geometries: influence of the main radius, of the pole gap height, of the angle of the sector (*RH* being constant).

2. Case of complex machines (cascades).

These are Mag + Electrostatic; as in Saclay
Mag + Mag; as in Orsay

The triple cascade — Mag + Mag + Electrostatic does not yet exist.

The analysis can be extended to this category of machines. The regimes of pressure are not the same as in the case of simple separators. As to the geometries it is found that small radii are in all respects better than large ones. This is not always true in the case of simple machines.

The enhancements of these machines are extremely high, higher than the product of multiplication of each magnetic stage, in the second case (Mag + Mag). [Cf. note added in proof.]

VII. Final Remark and Conclusion

All the perspectives are very promising but before concluding I wish to give some information which is likely to render the former perspectives obsolete.

Let us consider the hash contamination and the random displacement of the focus point. Let us assume that the hash level is just sufficient to contaminate the close neighbouring mass. The contamination beam is a pencil the direction of which is outside the frontier of the main beam. It follows that such contamination pencils can be stopped, and the corresponding contamination suppressed by proper use of aperture limiting shields without any intensity loss of the beam of the good isotope.

This fact which I have just deduced from the general hash manifestations, was actually discovered by chance. It is the subject of one above mentioned paper. I think that Dr. BERNAS will say some words about the experiment.

Some conclusions can be drawn. At first, the general theory can be upset in certain important and favourable points concerning the final enhancement factors, the working pressures, and the regimes of the sources. But this remark can immediately be put into practice.

Let us consider the case of a conventional instrument, e.g. a calutron, enhancing U^{235}. In order to minimize the hash contamination by U^{238}, one can try to use aperture screens before the collector entrance slit. After the preceding remark you can see that this screen must be located on the light masses side which seems paradoxical at a first glance. On the other hand, such a screen on the heavy side would improve the enhancement in respect of U^{234}.

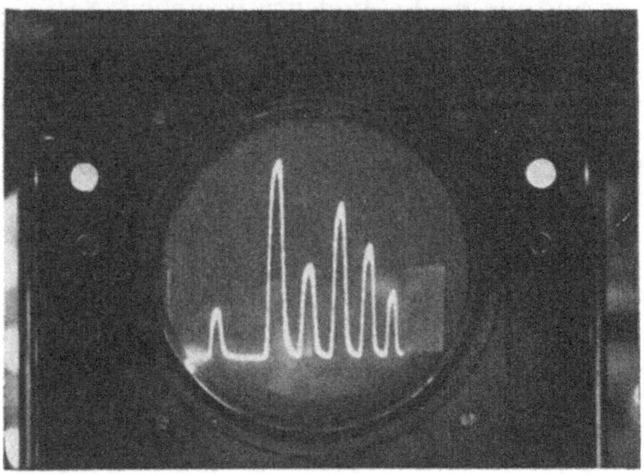

Fig. 11. Mass spectrum of mercury obtained with the two stages separator of Saclay, without aperture limiting shields

With the two stages separator at Saclay one can set shields about 10 cm behind the intermediate slit. Mass spectra are improved in shape when the positions of the screens are properly adjusted (Fig. 11, 12).

A recent collection of mercury 201 with such aperture limiters led to a final abundancy above 99.5%, whereas we could hardly reach 90% in former separations. The improvement lies mainly in respect

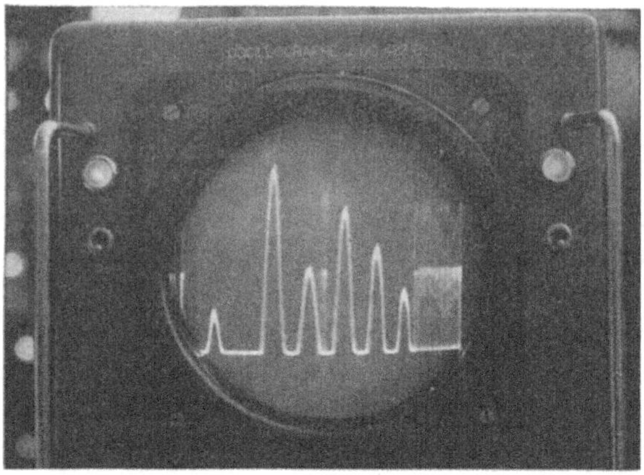

Fig. 12. The same mass spectrum, under the same conditions but with aperture limiting shields

of the isotopes 200 and 202, the closest neighbours. Mercury 196 was also processed and the final abundancy was about 83%.

These results can be considered as an experimental confirmation of the general theory and underline the future improvements which must be achieved, together with the desirability of controlling, corroborating, and improving the theory itself. But for this purpose the Saclay separator is not a suitable instrument.

Note added in proof.

As announced in the discussion, on page 80 some measurements of energy spread were recently performed, with an improved accuracy. An energy straggling probably due to scattering of the ions by neutrals at the beginning of the acceleration section was observed. The corresponding contamination seems to exist only for heavier isotopes. Its variation is not very simple and the values of the contamination factors can reach about some 10^{-4}.

Consequently this type of isotopic contamination can still be neglected in simple instruments and in cascades the second stage of which is electrostatic. On the other hand, the above conclusions about the two magnetic stage instruments no longer apply.

Discussion

J. Kistemaker: You mentioned in your paper an enrichment factor of 3000 for Hg^{196}. What was the collector current in this case?

Ch. Cassignol: We collected 0.3 mg within 60 hours.

J. Kistemaker: Have you tried to improve the stability of the beam against hash space charge disturbance by addition of electronegative gases e.g. O^-, Cl^-.

Ch. Cassignol: We intended to but have not yet done it. The best results were obtained with a very pure mercury metal.

J. Kistemaker: In connection with hash effects you referred to pressures of 2.10^{-4} and 5.10^{-5} mm Hg. Was this the pressure in the vacuum chamber or in the ion source?

Ch. Cassignol: In the vacuum chamber. This was a relative high pressure but the source was hashy. Generally the beam bursts at lower pressure but it also bursts at higher pressure when the source is hashy. Normally one cannot detect it then because of scattering effects but with the 2nd stage of the Saclay separator one can easily eliminate the scattering and see the beam burst.

P. Weinzierl: The mechanism of space charge compensation should be different within the magnetic field and outside of it. The machine of Dr. Cassignol contains straight parts outside the magnetic field. What can be said about the space charge compensation in this regard?

Ch. Cassignol: We have no information available so far.

W. Walcher: The mechanism of space charge compensation in a region with a magnetic field is not very different from the mechanism in a region without a field. The fast primary ions produce electrons and slow ions by collision with the residual gas. The slow ions are repelled by the positive space charge of the fast ions and, therefore, removed from the beam. Part of the electrons with thermal velocities, on the other hand, is captured by the beam and forms a neutralizing electron gas. Thus a flat potential valley is formed so that the boundary of the beam is a few volts less positive than its axis. From this potential valley thermal electrons escape by diffusion in such a quantity that the equilibrium of charge carriers is maintained. In a region without a magnetic field there is a diffusion of electrons from the beam in all directions perpendicular to the axis of the beam, in a region with strong magnetic field the diffusion across the field is hindered and the loss of electrons is only in the direction of the magnetic field. The drift motion, caused by the crossed electric and magnetic fields (a few volts between beam axis and beam boundary) may be neglected because the drift distance is very small.

R. Bernas: The space charge field within the analyser of the sector machine does not seem to me sufficiently high to explain the predominating contamination by heavier masses. In calutrons it would be more probable.

CH. CASSIGNOL: An extra proof for the proposed mechanism will be given now. We observed in the Saclay separator that the sign of the coma defect ($r\alpha^2$) changed when the intensity of the beam was raised. If the space charge potential inside the beam varies then the central part of the beam is more or less slowed down in respect to the rest of the beam and this can result into an abnormal coma defect when the intensity is high.

R. BERNAS: In Fig. 2 of DR. ALVÄGER's paper the predominating contamination is caused by light isotopes. Was it found to be generally true in all separations or was it valid only for this particular result?

T. ALVÄGER: A careful study of the intensity distribution was made only in the case of Kr^{85} and I^{131} but it seems to give about the same shape for other isotopes.

J. H. FREEMAN: We observed in our investigations an asymmetry in contamination of one isotope by another which is 8 masses away. This would require a very high space charge voltage. Does DR. CASSIGNOL believe that such high space charge voltage exists in fact under normal separating conditions or should this effect come from hash in the source unit?

CH. CASSIGNOL: There are no absolute values given in my paper.

R. BERNAS: As a starting hypothesis for his work DR. CASSIGNOL assumed that breakdowns did not occur frequently enough to be of any significance in contaminating neighbouring isotopes. However DR. FREEMAN has stated that breakdowns did occur frequently in his machine with some elements so that long range contamination could be explained by this process.

Adequate Characterization of Separating Power

By
Karl Ove Nielsen
Danish Atomic Energy Commission, Risö

With 2 Figures

The ultimate performance of an electromagnetic isotope separator depends on the materials to be separated and on the adjustment of the instruments. Therefore, it is difficult to describe an isotope separator by means of parameters characteristic of the instrument only.

Various possibilities of characterizing the separating power of an isotope separator will now be discussed.

At first, the separating power may be expressed by the resolving power (RP), which for laboratory separators is often defined by:

$$RP = \frac{D_1}{b_{1/2}} \cdot M \tag{1}$$

where

D_1 is the mass dispersion per mass unit,
$b_{1/2}$ is the width at half maximum (see Fig. 1),
M is the mass number of the isotope.

However, for $D_1 \gg b_{1/2}$ the resolving power does not give information about the completeness of the isotope separation. In this region, the resolving power is more an expression for the focusing conditions.

Fig. 1 shows the profile of a line corresponding to an isotope with the mass number M. The line is characterized by the width at half maximum ($b_{1/2}$), the "wings" (W) and the background (t). Fig. 1 also shows the slit with a width of S mm. In direct deposition defining slits normally are not used for restriction of the ion beams, and in such case S corresponds to the width of the employed part of the foil (Fig. 1).

The profile of a line is determined by the focusing conditions and the scattering of fast ions by collision with residual gas atoms in the vacuum system. While S and W mostly depend on the focusing conditions, it is the scattering by collision that is responsible for the background t. The background is therefore taken as

$$t = \alpha \cdot \mathcal{J} \tag{2}$$

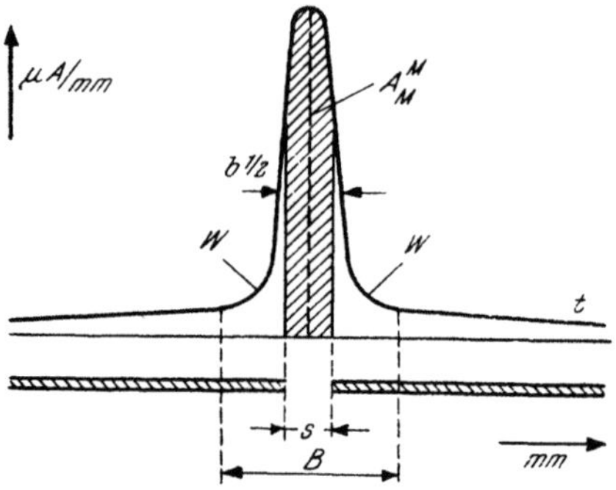

Fig. 1

where

$\mathcal{J}$ is the ion current causing the background,

α is a coefficient that depends not only on pressure and type of ions but also on the distance along the collection plate from the line to the place where t is measured.

By collection of isotopes with the mass number M, the background originating from isotopes with mass numbers different from M will cause contamination. The value t of the total background in the position of M may be expressed by [1]

$$\begin{aligned} t_M &= \alpha_1 \cdot \mathcal{J}^{M+1} + \alpha_2 \cdot \mathcal{J}^{M+2} + \cdots + \alpha_{-1} \cdot \mathcal{J}^{M-1} + \cdots = \\ &(\alpha_1 \cdot a_0^{M+1} + \cdots + \alpha_{-1} \cdot a_0^{M-1} + \cdots) \cdot \mathcal{J}_0 \ \mu\text{A/mm} \end{aligned} \tag{3}$$

where

$\mathcal{J}^{M+n}$ is the beam current corresponding to mass number $M+n$,

a_0^{M+n} is the original relative abundance of isotope with mass number $M+n$,

$\mathcal{J}_0$ is the total ion current,

α_n is the value of α at a distance corresponding to n mass units from M's position.

Through the slit (S) at M's position the following amounts of material will pass per second:

$$A_M^M = a_0^M \cdot \mathcal{J}_0 \cdot \frac{S}{b_{1/2}} \cdot C \tag{4}$$

$$A_c = t_M \cdot S \tag{5}$$

where

A_M^M is the amount of material corresponding to the mass number M at the position of M,

A_c is the total contamination originating from other isotopes deposited together with A_M^M,

C is a factor dependent on the ratio $S/b_{1/2}$ and on the shape of the line.
In the following, C is indicated for various values of $S/b_{1/2}$ [2].

Table

$S/b_{1/2} \ll 1$	$C \approx 1$
$S/b_{1/2} = 1$	$C \approx 3/4$
$S = B$	$C \approx 1/2$
$S > B$	$C \approx B/2S$

B is shown in Fig. 1.

The final relative abundance (a_f^M) of isotopes with the mass number M deposited on M's position is:

$$a_f^M = \frac{A_M^M}{A_M^M + A_c} \tag{6}$$

Correspondingly, the contamination of the deposited material on M's position is:

$$q = \frac{A_c}{A_M^M + A_c} \tag{7}$$

The enhancement E is defined by:

$$E = \frac{a_f^M}{1 - a_f^M} \cdot \frac{1 - a_0^M}{a_f^M} = \frac{1-q}{q} \cdot \frac{1 - a_0^M}{a_0^M} \tag{8}$$

Thus, inserting (4), (5) and (6) in eq. (8), we find [1]

$$E = \frac{C}{b_{1/2}} \cdot \frac{a_0^{M+1} + a_0^{M+2} + \cdots + a_0^{M-1} + \cdots}{\alpha_1 \cdot a_0^{M+1} + \cdots + \alpha_{-1} \cdot a_0^{M-1} + \cdots} \tag{9}$$

From eq. (9) it appears that although E depends on the adjustment of the instrument, E also depends on the relative abundance of the element. Therefore, the enhancement E is not very suitable for describing the performance of the separator.

From (9) it appears further that the less the α's deviate from each other, the less dependent is E on the mass spectrum. In the hypothetical case where $\alpha_1 \approx \alpha_2 \approx \alpha_3 \approx \cdots \approx \alpha_n \approx \alpha$ the enhancement factor is reduced to:

$$E \approx \frac{C}{b_{1/2}} \cdot \frac{1}{\alpha} \tag{10}$$

In this case, E is completely independent of the relative abundances.

Although in practice, the conditions for application of eq. (10) seldom exist, E is often used as a parameter describing the performance of the instrument. Thus we find directly from (8):

$$a_f^M = \frac{E \cdot a_0^M}{(E-1)a_0^M + 1} \quad \text{and} \quad q = \frac{1 - a_0^M}{(E \div 1)\, a_0^M + 1} \qquad (11)$$

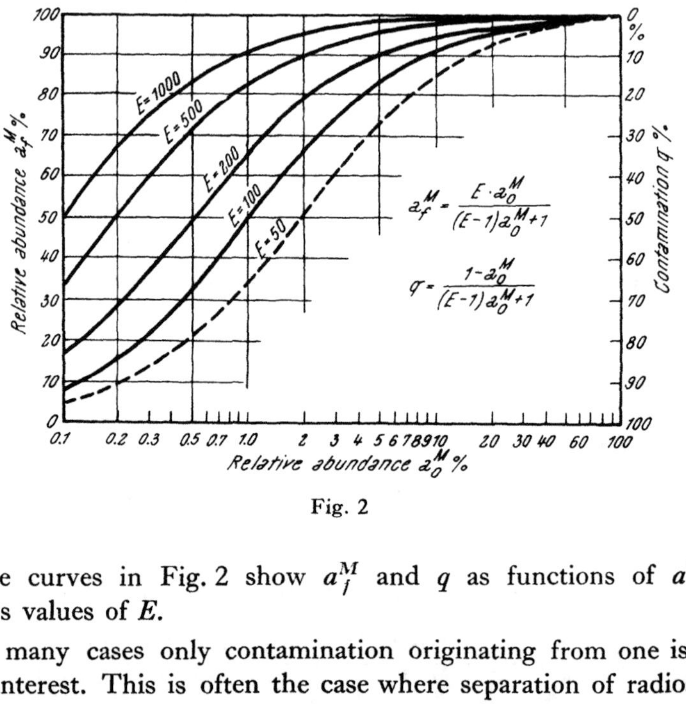

Fig. 2

The curves in Fig. 2 show a_f^M and q as functions of a_0^M for various values of E.

In many cases only contamination originating from one isotope is of interest. This is often the case where separation of radioactive isotopes is concerned. In such a case it is adequate to introduce the relative enhancement (E_R), see FREEMAN [3]. The relative enhancement factor ${}^M E_R^{M+n}$ of the isotope with mass number M with respect to the isotope with mass number $M+n$ is defined by:

$${}^M E_R^{M+n} = \frac{a_f^M}{a_f^{M+n}} \cdot \frac{a_0^{M+n}}{a_0^M} \qquad (12)$$

Substituting (4) and (5) in eq. (12), we find:

$${}^M E_R^{M+n} = \frac{C}{b_{1/2}} \cdot \frac{1}{\alpha_n} \qquad (13)$$

From eq. (13) it appears that the relative enhancement is independent of the mass spectrum of the element, provided that lines corresponding to different mass numbers have the same shape. Therefore, the relative enhancement (E_R) is more suitable than E for characterizing the separating power of an instrument.

Since the different beam currents are unlike, the current distribution will not have the same shape. Further, the ion current distribution of a line is not symmetrical ($a_n \neq a_{-n}$). Thus, E_R is not constant "over" the mass lines of an element, although the variations will be considerably smaller than for the enhancement factor E.

Both E and E_R depend on S and $b_{1/2}$. From (9) and (12) and the Table it appears that:

1. E and E_R are inversely proportional to $b_{1/2}$ when $S/b_{1/2}$ is constant.
2. E and E_R increase with decreasing S when $b_{1/2}$ is constant. The increase of E and E_R takes place at the expense of efficiency.
3. For $S \ll b_{1/2}$, E and E_R are independent of S and inversely proportional to $b_{1/2}$.

As appears from (12), ${}^{M}E_R^{M+n}$ is measured by determining the ratio a_f^M / a_f^{M+n}, which is based on a mass analysis of the material deposited on M's position. If a_n is determined from (13) for a series of values of n, it is possible to fix the ion current distribution for a mass line [3].

Since both E and E_R depend on S and $b_{1/2}$, which are often altered, the separating power of the instrument may be more directly expressed by the distribution curve (a).

References

1. Nielsen, K. O., Thesis, 1958.
2. Nielsen, K. O., and O. Skilbreid, Nuclear Instruments II, 15 (1958).
3. Freeman, J. H., Symposium on "Electromagnetic Separation of Radioactive Isotopes" Vienna, 23rd-25th May, 1960, present volume, Chapter I. p. 40 ff.

The Relative Enhancement Factor in Electromagnetic Isotope Separation

By

J. H. Freeman

Chemistry Division, A. E. R. E., Harwell

With 14 Figures

Summary

In part 1 a relative enhancement coefficient which describes the performance of an electromagnetic isotope separator is defined. This coefficient is then used to calculate the resolution of the isotopic beams, and a method of predicting enhancements and detecting mass analysis errors is given. Part 2 gives a brief review of the causes of isotopic contamination and demonstrates how the calculated beam shape can be used to evaluate the importance of the various causes.

Introduction

The electromagnetic method of isotope separation is unique in that a considerable enhancement of all the isotopes of any polyisotopic element may be obtained simultaneously and in a single stage separation. Although under ideal conditions virtually complete resolution of the separate ion beams may be obtained (Fig. 1), it is found in practice that the collected isotopes are never completely pure. In order to improve separator performances it is important to understand the causes of isotopic contamination and to have a means of estimating their separate contributions to the total loss of resolution. In the case of radioactive separations it is also important to be able to predict, or calculate, from available data the purity of an isotope which may be too active, or of too low abundance, to allow an analysis.

The performance of a separator is generally described by the isotopic enhancement factor

$$E = \frac{C_f}{100 - C_f} \cdot \frac{100 - C_o}{C_o}$$

obtained for a given separation. Since the enhancement so defined is a function not only of the resolving power of the separator, but also of the original abundance of all the neighbouring isotopes the value of this concept is limited to comparisons of separations of the same source material. It will give no measure of the causes of contamination

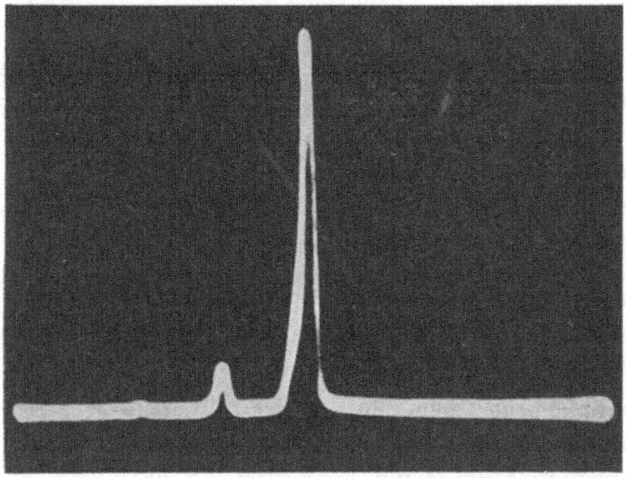

Fig. 1. Oscilloscope trace of plutonium beam

and cannot be used to compare separations of isotopes of different elements, nor separations of the same element using source materials of varying abundances.

It is possible to define a relative enhancement factor (viz. the enhancement of one isotope with respect to another) which is found to be a measure only of the performance of the separator, and which can provide the basis of a sensitive method of studying and comparing the behaviour of electromagnetic separators.

Part 1

The Relative Enhancement Factor and the Beam Spread

Definition

The isotopic enhancement is normally defined as

$$E = \frac{Cf}{(100 - Cf)} \cdot \frac{(100 - Co)}{Co}$$

where Co and Cf are the original and final concentrations of an isotope.

We can define the relative enhancement factor E_R for a specific case as

$$^{a}E_R{}^{b} = \frac{Cf_a{}^{a}}{Cf_a{}^{b}} \cdot \frac{Co_b}{Co_a} \tag{2}$$

where $^{a}E_R{}^{b}$ is the enhancement of isotope a with respect to b; $Cf_a{}^{a}$ and $Cf_a{}^{b}$ are the final abundances of isotopes a and b in collector a; and Co_a and Co_b are the abundances of a and b in the source material.

If equation 1 is rewritten $\frac{Cf_a{}^{a}}{Cf_a{}^{b}} \Big| \frac{Co_a}{Co_b}$ it can readily be seen that $^{a}E_R{}^{b}$ is simply the ratio of the abundances of isotopes a and b in the collector a to their abundances in the source material. For a two component system $E = E_R$.

For simplicity it is generally simpler to write $^{a}E_R{}^{b}$ as $^{a}E^{\Delta n}$, or even $E^{\Delta n}$ if the mass number is understood, e.g. the enhancement of Pu^{239} with respect to Pu^{242} could be written $^{Pu^{239}}E_R{}^{Pu^{242}}$, $^{Pu^{239}}E^{+3}$ or E^{+3} and the enhancement of Li^7 with respect to Li^6 could be written $^{Li^7}E^{Li^6}$, $^{Li^7}E^{-1}$, $^{7}E^{-1}$ or simply E^{-1}.

The concept of a relative enhancement factor as described above is not novel. It has been used by Bernas [1] and Livingston [2] to obtain more meaningful descriptions of separator performance and at A.E.R.E. to predict likely results of separations on Hermes using plutonium source materials of varied abundances. It is similar in form to the "relative volatility factor" [3] used in fractional distillation and the "separation factor" [4] used in chemical separation processes.

It has not however been adopted generally in discussions of isotope enhancement, nor have its possibilities as a tool for studying separator performance been fully exploited.

Comparison of the Two Enhancement Factors

In the case of separations involving only two isotopes the enhancement factors E and E_R are identical, but this is not the case for elements naving more than two isotopes. Table 1 compares E and E_R for a heodymium separation.

Table 1. *Enhancement factors for Neodymium Separation Run E.Q.: O.R.N.L.* [5]

Enhancement Factor		Isotope							
	Δn	Nd^{142}	Nd^{143}	Nd^{144}	Nd^{145}	Nd^{146}	Nd^{148}	Nd^{150}	mean
E		35	38	46	41	116	130	306	
E_R	+1	15	19	14	12				15 ± 4
E_R	−1		46	51	47	45			47 ± 4
E_R	+2	30	32	29		22	23		27 ± 5
E_R	−2			151	138	147	110	77	125 ± 48

The relative enhancement factors of constant 'Δn' for the above separations are reasonably constant whereas the ordinary enhancement factors vary widely from one isotope to another. It will be shown later that the small differences in the values of E_R can give valuable information on the causes of contamination.

Calculation of the Isotope Distribution at the Collector (Beam Spread)

The beam spread of an isotope is its intensity distribution across the whole of the collector assembly. That is to say it is a measure of the rate of arrival of *both* neutral and charged species.

During the course of a separation the ion beams can be monitored by means of a probe and suitable adjustments made to the source to obtain the required degree of resolution, but no measurement can be made of the neutral particles which simultaneously deposit in the isotope receivers. Thus although under ideal conditions complete resolution of the ion beams may be obtained the collected isotopes are never 100% pure.

If all of the data of an isotope collection were available the beam spread of each isotope could be exactly calculated.

The beam spreads can however be calculated from the mass analysis alone if it is assumed (a) that the ratio of the rates of arrival of any two isotopes in one pocket is given by their ratio in the mass analysis of that pocket (i.e. all losses are proportional to the abundance), and (b) that the ratio of the rates of arrival of any isotopes in their own pockets is the same as the ratio of their natural abundances.

This calculation requires a complete knowledge of the mass analyses, but in fact if it is assumed that all of the beams are equally resolved and have the same shape then it is possible to calculate the beam shape from the mass analyses of only one or two isotopes.

The beam shape so obtained can be used for the prediction and checking of mass analyses and for the evaluation of collector behaviour.

(i) *Complete calculation*

Although a general treatment can readily be developed, it is less confusing to treat a particular example. Consider the neodymium separation (run E. Q. O. R. N. L. [5]) for which complete analyses are given below in Table 2.

Table 2. *Mass analyses for Neodymium Separation E. Q.: O. R. N. L.*

Isotope	Sample	Atom % in Sample						
		142	143	144	145	146	148	150
142	587 (a)	**93.69**	2.88	2.77	0.29	0.33	0.03	0.02
143	588 (a)	4.04	**83.93**	8.83	1.78	1.16	0.15	0.108
144	589 (a)	0.698	0.941	**93.45**	2.33	2.36	0.147	0.066
145	590 (a)	1.18	0.84	4.83	**78.60**	13.71	0.626	0.214
146	591 (a)	0.584	0.274	0.903	1.03	**95.60**	1.42	0.192
148	592 (a)	1.21	0.584	1.28	0.74	2.46	**89.85**	3.88
150	593 (a)	1.10	0.456	0.98	0.419	1.04	1.25	**94.76**

In order to calculate the beam spread of say the Nd^{142} isotope across the collector assembly we need to know the ratio of its rate of arrival in the 142 collector to its rate of arrival in each of the other collectors. This can be calculated als follows:

$$\frac{\text{Rate of arrival of 143 in 143 collector}}{\text{Rate of arrival of 142 in 142 collector}} = \frac{Co_{143}}{Co_{142}} \tag{3}$$

Also

$$\frac{\text{mean rate of arrival of 142 in 143 collector}}{\text{mean rate of arrival of 143 in 143 collector}} = \frac{Cf_{143}^{142}}{Cf_{143}^{143}} \tag{4}$$

whereas previously defined $Cf_b{}^a$ is the final abundance of isotope a in collector b.

3 and 4 give

$$\frac{\text{mean rate of arrival of 142 in 143 collector}}{\text{mean rate of arrival of 142 in 142 collector}} = \frac{Cf_{143}^{142}}{Cf_{143}^{143}} \cdot \frac{Co_{143}}{Co_{142}} \tag{5}$$

$$\text{but by definition (eq. 2)} \quad \frac{Cf_{143}^{142}}{Cf_{143}^{143}} \cdot \frac{Co_{143}}{Co_{142}} = \frac{1}{{}^{143}E^{142}} \tag{6}$$

5 and 6 give

$$\frac{\text{mean rate of arrival of 142 in 143 collector}}{\text{mean rate of arrival of 142 in 142 collector}} = \frac{1}{{}^{143}E^{142}}$$

Similarly

$$\frac{\text{mean rate of arrival of 142 in 144 collector}}{\text{mean rate of arrival of 142 in 142 collector}} = \frac{1}{{}^{144}E^{142}} \quad \text{etc.}$$

The ratio of the rate of arrival of Nd^{142} in its own collector to its rate of arrival in the other collectors can thus be calculated and its beam spread can be expressed as a series of ratios as follows:

$$1 : \frac{1}{{}^{143}E^{142}} : \frac{1}{{}^{144}E^{142}} : \frac{1}{{}^{145}E^{142}} : \frac{1}{{}^{146}E^{142}} \quad \text{etc.} \tag{7}$$

or more conveniently multiplying 7 throughout by 100

$$100 : \frac{100}{{}^{143}E^{142}} : \frac{100}{{}^{144}E^{142}} : \frac{100}{{}^{145}E^{142}} : \frac{100}{{}^{146}E^{142}} \quad \text{etc.} \tag{7a}$$

(7a) can be written:

$$100 : \frac{100}{E^{-1}} : \frac{100}{E^{-2}} : \frac{100}{E^{-3}} : \frac{100}{E^{-4}} \quad \text{etc.} \tag{8}$$

Similar calculations for each of the other isotopes will give equations similar to eq. (8) from which it will be seen that $E^{\Delta n}$ is substantially the same for each value of Δn irrespective of which isotopes are considered. A plot can be made of all the values obtained from eq. (8) when it is found that each equation gives points on a *part* of an overall curve describing the isotope beam distribution. It should be noted that *positive* values of Δn give the distribution on the *lower* mass part of the curve and that *negative* Δn values give the *higher* mass part.

(ii) *Simplified calculation*

If it is assumed that the isotope beams are equally resolved it is possible to make the analyses in terms of the separate collectors rather than the separate isotopes.

For example, consider only the mass analyses of the collectors 142 and 150 of the neodymium separation quoted above (Table 2). From these we can calculate:

$$100:\frac{100}{{}^{142}E^{143}}:\frac{100}{{}^{142}E^{144}}:\frac{100}{{}^{142}E^{145}}:\frac{100}{{}^{142}E^{146}}:\frac{100}{{}^{142}E^{148}}:\frac{100}{{}^{142}E^{150}} \tag{9}$$

and

$$\frac{100}{{}^{150}E^{142}}:\frac{100}{{}^{150}E^{143}}:\frac{100}{{}^{150}E^{144}}:\frac{100}{{}^{150}E^{145}}:\frac{100}{{}^{150}E^{146}}:\frac{100}{{}^{150}E^{147}}:100 \tag{10}$$

respectively.

If ${}^{x}E^{\Delta n}={}^{y}E^{\Delta n}=E^{\Delta n}$ etc. then (9) and (10) can be written:

$$100:\frac{100}{E^{+1}}:\frac{100}{E^{+2}}:\frac{100}{E^{+3}}:\frac{100}{E^{+4}}:\frac{100}{E^{+5}}:\frac{100}{E^{+8}} \tag{11}$$

and

$$\frac{100}{E^{-8}}:\frac{100}{E^{-7}}:\frac{100}{E^{-6}}:\frac{100}{E^{-5}}:\frac{100}{E^{-4}}:\frac{100}{E^{-2}}:100 \tag{12}$$

respectively, hence the complete distribution:

Mass units	. . . −3	−2	−1	0	+1	+2	+3 . . .
Distribution	. . . $\frac{100}{E^{+3}}$	$\frac{100}{E^{+2}}$	$\frac{100}{E^{+1}}$	100	$\frac{100}{E^{-1}}$	$\frac{100}{E^{-2}}$	$\frac{100}{E^{-3}}$ . . .

can be obtained.

(iii) *Prediction*

The distribution obtained in the manner described above can be used to predict the enhancements of either unanalysed samples using the available but incomplete mass analysis results or to predict the likely enhancements of future separations.

For example the distribution of the Nd^{143} could be calculated from the beam distribution (eq. 13) obtained from the 142 and 150 results as follows.

For every 100 Nd^{143} atoms in the 143 pocket there will be:

$$
\begin{aligned}
&\frac{100}{E^{+1}}\,\frac{Co_{142}}{Co_{143}} \text{ atoms of } 142\\
&\frac{100}{E^{-1}}\,\frac{Co_{144}}{Co_{143}} \text{ atoms of } 144\\
&\frac{100}{E^{-2}}\,\frac{Co_{145}}{Co_{143}} \text{ atoms of } 145\\
&\frac{100}{E^{-3}}\,\frac{Co_{146}}{Co_{143}} \text{ atoms of } 146\\
&\frac{100}{E^{-5}}\,\frac{Co_{148}}{Co_{143}} \text{ atoms of } 148\\
&\frac{100}{E^{-7}}\,\frac{Co_{150}}{Co_{143}} \text{ atoms of } 150
\end{aligned}
\tag{13}
$$

Calculation of these ratios and converting to percentages will give the Nd^{143} mass analysis.

Similarly, to predict future separation results, using say source material of a different isotopic composition, the same calculation is made using the new values of *Co* for each isotope.

Results and Discussion of some Distribution Curves

Fig. 2 shows the composite distribution curve calculated as described above for all isotopes of the neodymium separation E.Q.—Table 2. Also shown are the curves obtained for the Nd^{142} and Nd^{150} collectors only. From these last two curves the distribution given in Table 3 can be obtained.

Table 3. *Neodymium distribution from the 142 and 150 Collectors only*

Mass Units	−8	−7	−6	−5	−4	−3	−2	−1	0
Distribution	0.115	0.13	0.2	0.35	0.55	1.01	3.33	7.62	100

Mass Units	0	+1	+2	+3	+4	+5	+6	+7	+8
Distribution	100	2.8	1.0	0.6	0.36	0.3	0.26	0.23	0.21

This distribution can now be used to predict the mass analyses of the other isotopes and Table 4 gives the mass analyses for the 143, 144, 145, 146 and 148 isotopes as measured and as calculated from the 142 and 150 pockets.

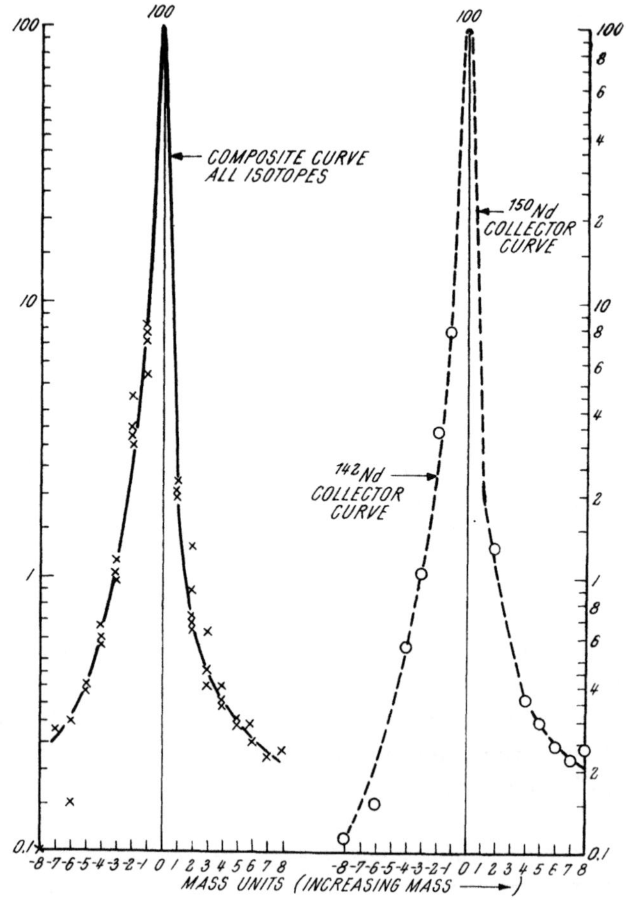

Fig. 2. Neodymium distribution (Run E. Q. O. R. N. L.)

Although Table 4 represents an extreme case in which only two analyses are used to calculate the remaining five the agreement between the measured and calculated abundances is very reasonable.

It is also possible, in principle at least, to use the distribution curve of one separation to predict the likely results of future separations. Such a calculation would only be valid if the operating conditions of the first separation were reproducible, and then only for source materials of similar mass and properties, such as the actinide or lanthanide series of elements, or for irradiated samples of the same element having varied isotopic compositions.

Table 4. *Measured and Calculated Mass Analyses*

Sample	Isotope						
	142	143	144	145	146	148	150
Nd^{143} measured	4.04	**83.93**	8.83	1.78	1.16	0.15	0.108
calculated	5.0	**80.1**	12.0	1.8	1.1	0.13	0.046
Nd^{144} measured	0.698	0.941	**93.45**	2.33	2.36	0.147	0.066
calculated	1.05	1.32	**92.8**	2.44	2.2	0.12	0.042
Nd^{145} measured	1.18	0.84	4.83	**78.6**	13.71	0.626	0.214
calculated	1.5	1.1	6.35	**77.9**	12.3	0.543	0.2
Nd^{146} measured	0.584	0.274	0.903	1.03	**95.60**	1.42	0.192
calculated	0.54	0.4	1.32	1.27	**95.5**	1.05	0.17
Nd^{148} measured	1.10	0.456	0.98	0.419	1.04	**89.85**	3.88
calculated	1.1	0.59	1.37	0.8	2.74	**90.7**	2.9

It follows that such calculations are unlikely to be accurate for high-output calutrons where conditions are rarely reproducible. For low-output, high-resolution machines the predictions would probably be sufficiently accurate to determine, for example, the necessary degree of irradiation of a source material to obtain the required purity of the separated isotopes.

The original separations in Hermes were carried out on low "burn-up" plutonium. It was desirable to predict the likely abundances which would be obtained using a much higher "burn-up" source material. Fig. 14 shows the distribution curve obtained for a separation (H. 30) using the first source material whose mass analysis was:

Isotope	239	240	241	242
Abundance	93%	6.48%	0.51%	0.028%

The curve gives the following mean distribution of the plutonium beam

Mass Units	—3	—2	—1	0	1	2	3
Distribution	0.32	0.5	1.0	100	0.54	0.18	0.1

For source material of mass analysis:

Isotope	239	240	241	242
Abundance	63.7%	30.9%	4.49%	0.97%

we can calculate the likely abundances; these are given in Table 5.

Table 5. *Predicted Plutonium Analyses*

Sample	Isotope			
	239	240	241	242
239	**99.5**	0.48	0.035	0.005
240	1.1	**98.8**	0.072	0.016
241 .	2.4	3.5	**93.9**	0.2
242	5.7	5.0	2.2	**87.1**

Table 6 gives analyses actually obtained for two separations using the enriched source material.

Table 6. *Measured Analyses*

Run B 5 C

Sample	Isotope			
	239	240	241	242
239	**99.0**	0.89	0.69	0.023
240	.664	**99.2**	.078	.016
241	2.54	2.68	**94.6**	0.18
242	8.74	7.21	1.14	**82.9**

B 7–9 E

Sample	Isotope			
	239	240	241	242
239	**99.5**	0.42	0.036	0.013
240	1.54	**98.3**	0.11	0.015
241	4.45	5.68	**89.6**	0.21
242	11.36	13.5	2.3	**72.9**

As can be seen from Tables 5 and 6 the agreement between the predicted and measured analyses is as good as the agreement between separate runs.

The two methods of calculating the beam spread described in the section above evidently give the same composite curve if the complete mass analysis results are used. Nevertheless the second method, viz. calculation by pockets, is generally more useful since it emphasises the differences in collector behaviour and clearly shows up any gross errors in the mass analyses. Fig. 4 for example shows the pocket distributions for the neodymium separation Run J. I. at O. R. N. L. [5]. As can be seen the 142, 146, 148 and 150 curves are not so well resolved as the 143, 144 and 145 curves. This is presumably due to the fact that the pocket dimensions were not identical.

Isotope	142	143	144	145	146	148	150
Internal pocket width	5/16″	3/16″	3/16″	3/16″	5/16″	1/4″	1/4″

In the neodymium collection E. Q. previously discussed, Fig. 5, the pockets were identical and the curves are all equally resolved.

Figs. 10 and 11 show the beam spreads obtained for an A. E. R. E. tellurium separation and an O. R. N. L. chromium separation [6]. In each figure it can be seen that one of the collector curves does not lie on the composite curve for the other collectors. Re-analysis [5] of the samples in question showed the original figures to be in error. The corrected analyses fall as would be expected on the composite curve.

The method may also be useful in demonstrating the presence of small quantities of isobaric impurities, e.g. only one per cent of samarium in gadolinium source material would lead to gross contamination of the low abundance (0.20%) Gd^{152} by the high abundance (26.8%) Sm^{152}. If, as is possible, the mass spectrometer did not discriminate between the two isobars the impurity could be readily detected by comparing the distribution curve for the 152 collector with the composite curve for the other collectors.

Finally a careful inspection of a large number of distribution curves for a variety of separations shows no consistent difference between the shapes of the curves for different isotopes of a given element. *It would therefore appear that under normal conditions all of the isotope beams of a given separation are equally resolved at the collector.*

Part 2

Isotopic Contamination in Electromagnetic Separation

Now that stocks of separated isotopes of almost all of the poly-isotopic elements are commercially available, attention is being focused on the achievement of higher purities and the separation of radioactive isotopes. It is becoming increasingly important to obtain an understanding of the causes of contamination and to find ways of minimising them. Some insight can be obtained into these problems by the application of the method described above.

Contamination may result from a large number of causes, the most important of these are listed below.

(a) Focusing defects of the magnetic analyser.
(b) Inadequate focusing at the source unit.
(c) Space charge repulsion effects.
(d) Gas collision without loss of charge (elastic collisions).
(e) Instability of source supplies and electrical breakdown (sparking).
(f) Gas collisions giving rise to charge neutralisation (inelastic collisions).
(g) Source material sputtering from walls of vacuum tank or vapourised directly from the source unit.
(h) Sputtering or evaporation of one receiver to another.

Isotope Contamination and the Beam Shape

The causes (a)–(e) listed above will all result in spread of the ion beam and are all therefore in principle capable of being detected and estimated by electrical measurements and at least partially corrected. For example Fig. 1 shows virtually complete resolution of the ion beam and it follows that any ionic contamination must result from cause (e) which is spasmodic or from cause (d) at a very low level.

The likely effects (grossly exaggerated) of these last two causes of contamination on the overall beam distribution are shown in Fig. 3 (d)

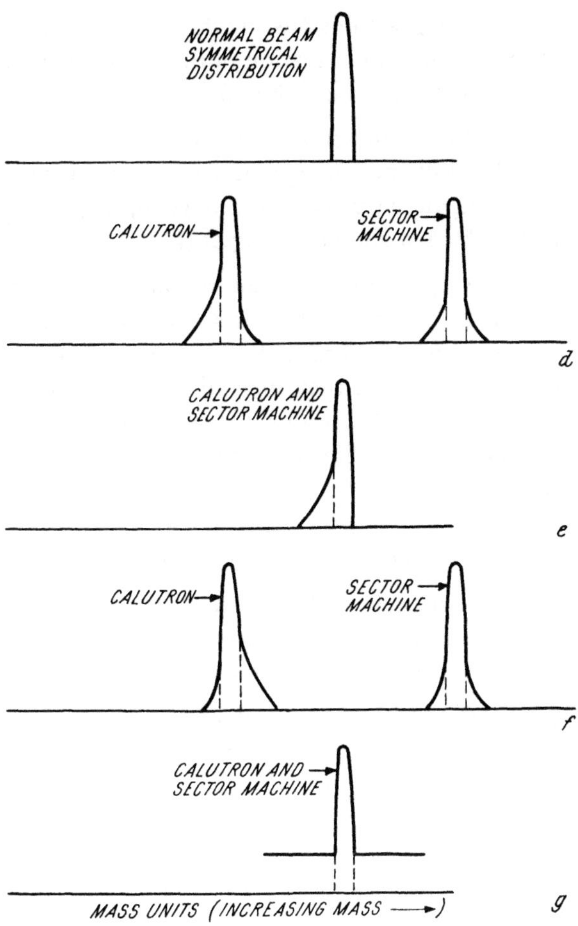

Fig. 3. Effects of contamination on beam distribution

and (e). Elastic gas collision will result in both deflection and loss of energy and in the calutron this will give rise to an asymmetric distribution. Electrical breakdown resulting in a loss of energy at the source unit will give an asymmetric distribution both in the calutron and in sector machines.

Contamination of the collectors by neutrals is not normally detectable during the course of a collection although in the case of certain radioactive separations such a measurement is feasible at least in principle.

Inelastic small angle gas collisions in which the ion loses its charge will result in asymmetric contamination in the calutron and symmetric contamination in a sector machine. Fig. 3 (f).

Condensation of source material in the receivers either by repeated sputtering from the walls of the separator or direct evaporation from the source unit (e.g. mercury, chlorine, oxygen, sulphur) will result in the type of distribution shown in Fig. 3 (g).

Contamination by sputtering or evaporation from one collector to another can be almost eliminated by careful design of the pocket assembly, but in separations where it does occur it is very difficult to predict the likely results since they depend upon the geometry of the collectors, the intensity of the beams and a number of other factors. It is likely however that the collectors closest to the abundant isotopes will be most affected.

Analysis of some Separations

The general self consistency of the calculated results suggests that they have real foundations and that they can be used to obtain information from the various beam shapes found in practice.

In normal separator operation, most of the likely causes of contamination will be contributing in varying degrees to the total loss of enhancement. Since only one composite curve can be obtained from the mass analysis, it is evident that all of the causes cannot be separately assessed. Nevertheless, a considerable amount of information can be obtained since different causes have a different effect on the curves; this, if necessary, should be amplified in a specific case by carefully measuring the mean ion beam distribution and correlating the two results.

A number of calculated distribution curves of typical separations carried out in O.R.N.L. or A.E.R.E. separators are shown below (Figs. 4–14) with comments on the likely causes of contamination.

Fig. 4. The double curve attributed to pocket size differences has already been discussed (p. 51). The curve at lower masses (L.H.S.) is considerably broader than at higher masses (R.H.S.). It seems

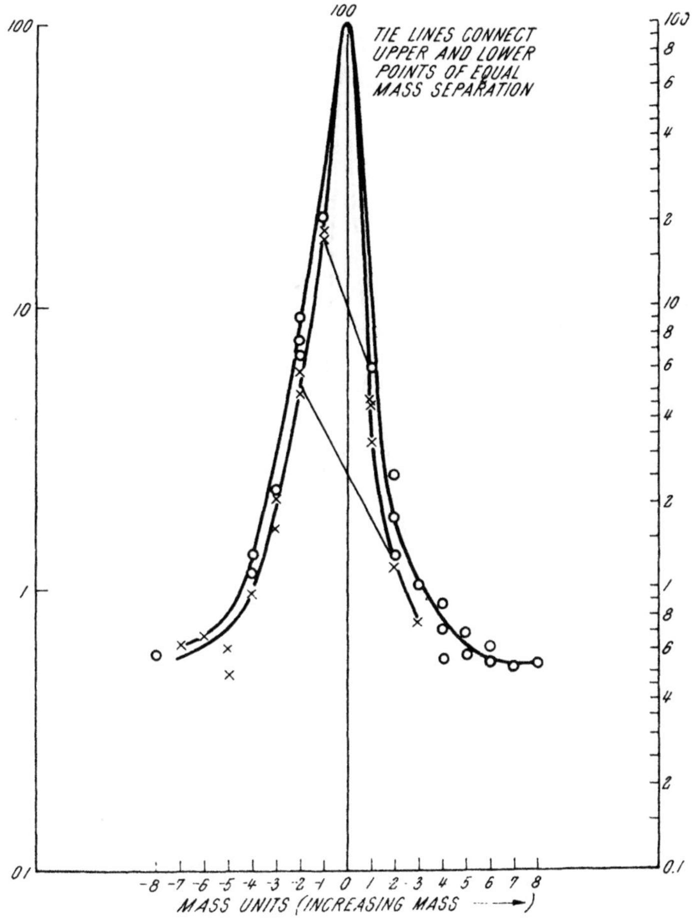

Fig. 4. Neodymium distribution (Run J. I. O. R. N. L.)

likely that this is due to sparking. Frequent sparking is a characteristic of the lanthanide elements, and in fact, in this case it appears to represent the principal limitation of enhancement. From the lower tails of the curve, we can see that the contamination due to source material [cause (g)], is less than 0.5, that is to say less than 0.5% of the total collection.

Fig. 5. In this neodymium collection, there is no sign of a double curve (the pockets are identical). Again, the asymmetry due to spark-

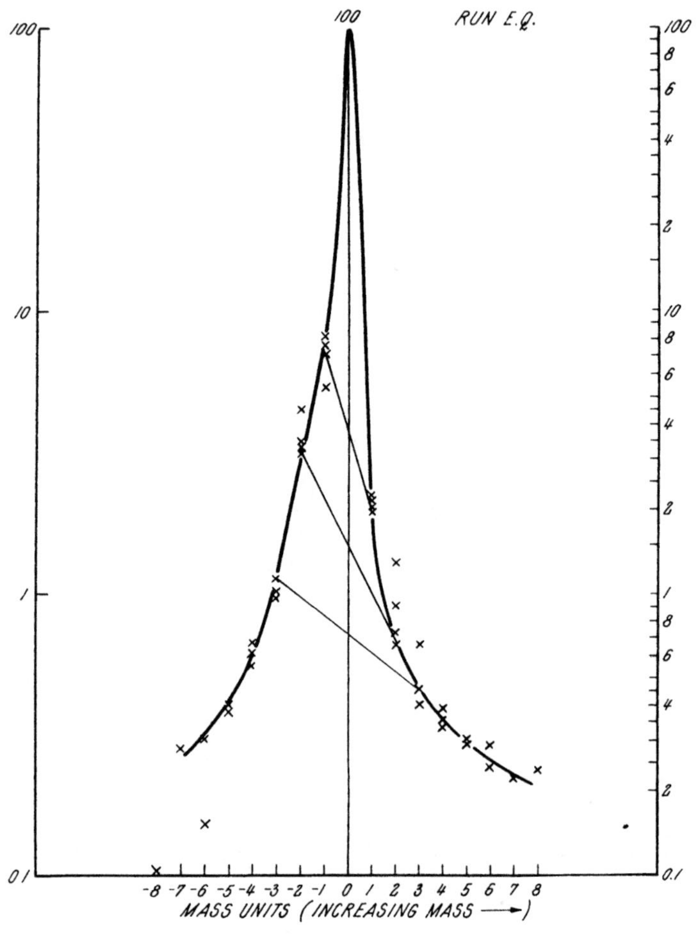

Fig. 5. Neodymium distribution (Run E. Q. O. R. N. L.)

ing can be readily observed, neutral contamination is less than 0.2% of the total collection. The sparking is not so severe as in Fig. 4, and the overall resolution is somewhat better.

Fig. 6. Shows an A.E.R.E. samarium separation. The sparking appears to be much less severe than in Figs. 4 and 5, although

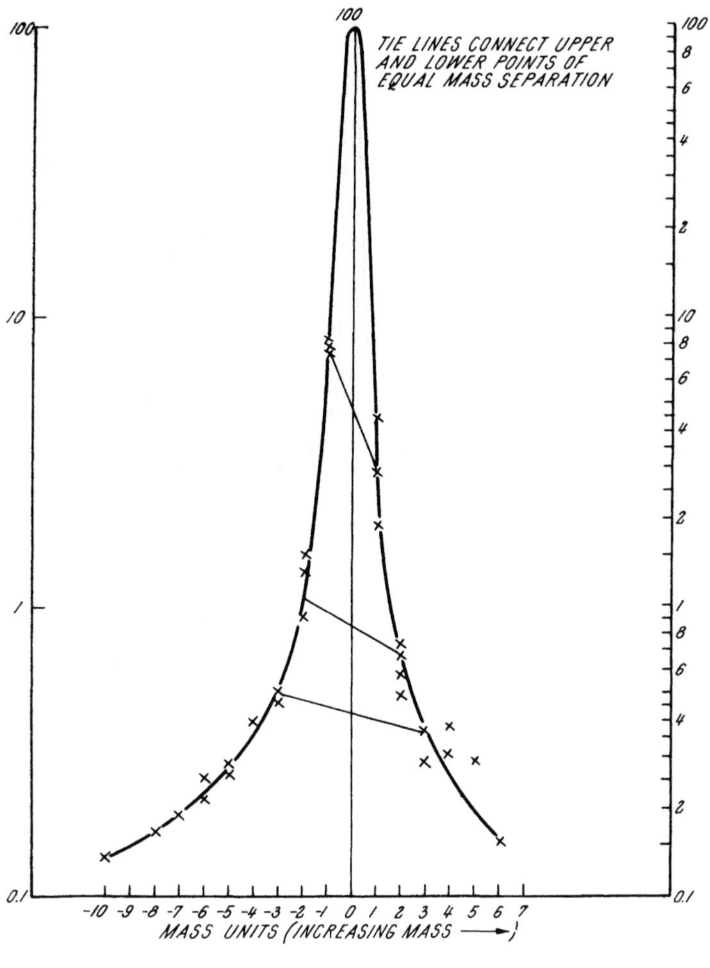

Fig. 6. Samarium distribution (A.E.R.E.)

it is still causing approximately half of the lower mass contamination. Source material contamination is less than 0.1% of the total collection.

Fig. 7. This is one of the most interesting cases to study, since it shows very clearly the effect of source material contamination [section (g)]. In spite of the very considerable precautions taken to condense

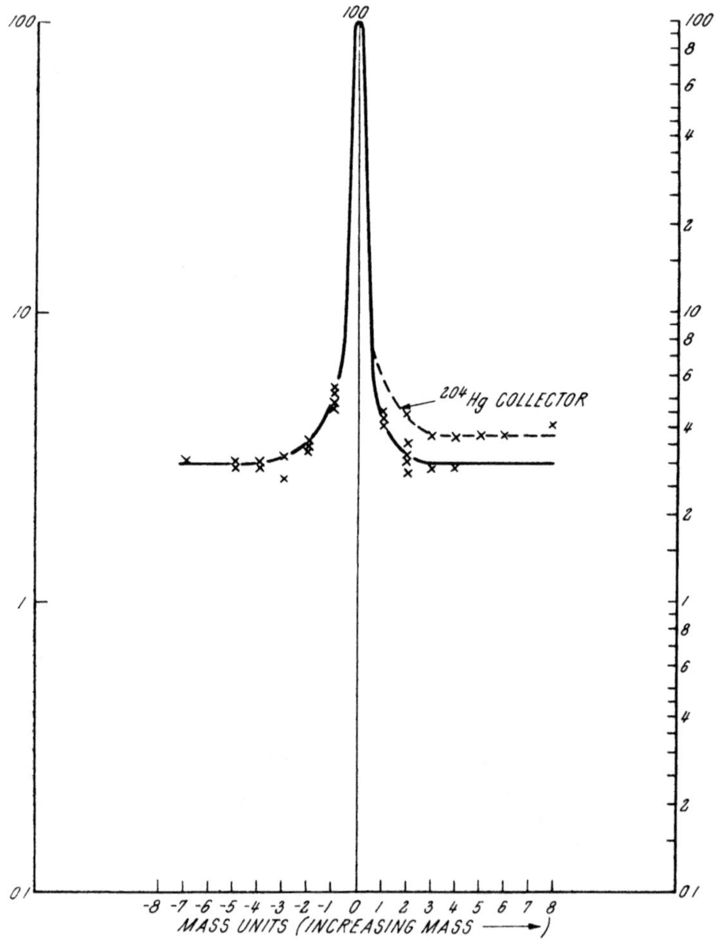

Fig. 7. Mercury distribution (O. R. N. L.)

the mercury vapour and prevent it reaching the collector unit 6 and 7, it can be seen that approximately 5 % of the total collection is due to unseparated mercury element. Sparking has had little effect, as would be expected. The scatter of results on the higher mass side is probably due to the fact that the pockets were processed sectionally (see Fig. 8).

The results plotted are the calculated mean analyses for all sections, and evidently incomplete recovery of any of the sections would result in the mean being weighted.

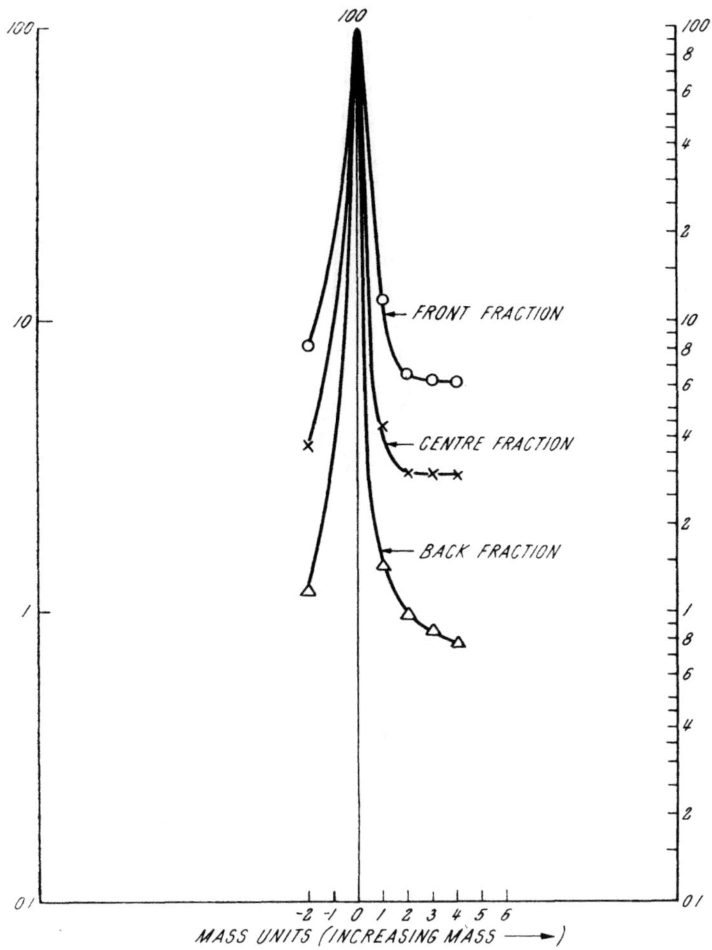

Fig. 8. Mercury distribution (202 collector only, O. R. N. L.)

Fig. 8. Shows the separate curves obtained for the three sections of the Hg^{202} collector [5], and clearly demonstrates the value that this technique may have in obtaining high purity samples.

Fig. 9. Shows the results obtained for zinc [8]. Here, as might be expected, the results show a considerable scatter, probably due to

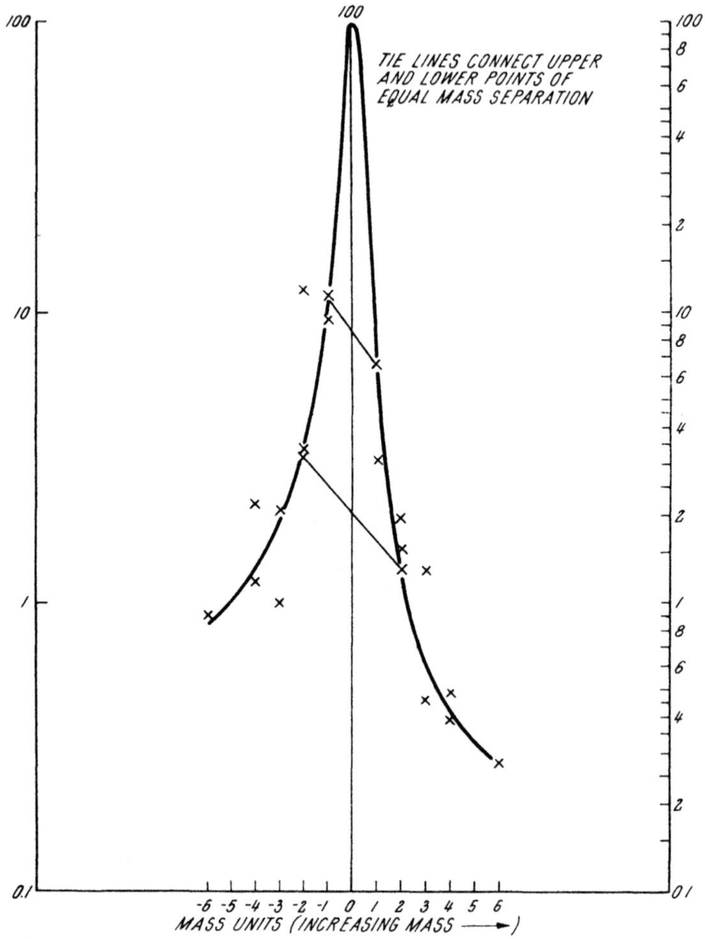

Fig. 9. Zinc distribution (O. R. N. L.)

sputtering and evaporation of the volatile zinc from one collector to another [section (h)]. Another interesting feature is that although zinc is notoriously volatile in vacuum systems and would be expected to give one of the highest values for source material contamination, this effect appears to be less than 0.3 % of the total collection.

Fig. 10. An almost symmetrical tellurium curve showing no significant sparking contamination. Source material contamination

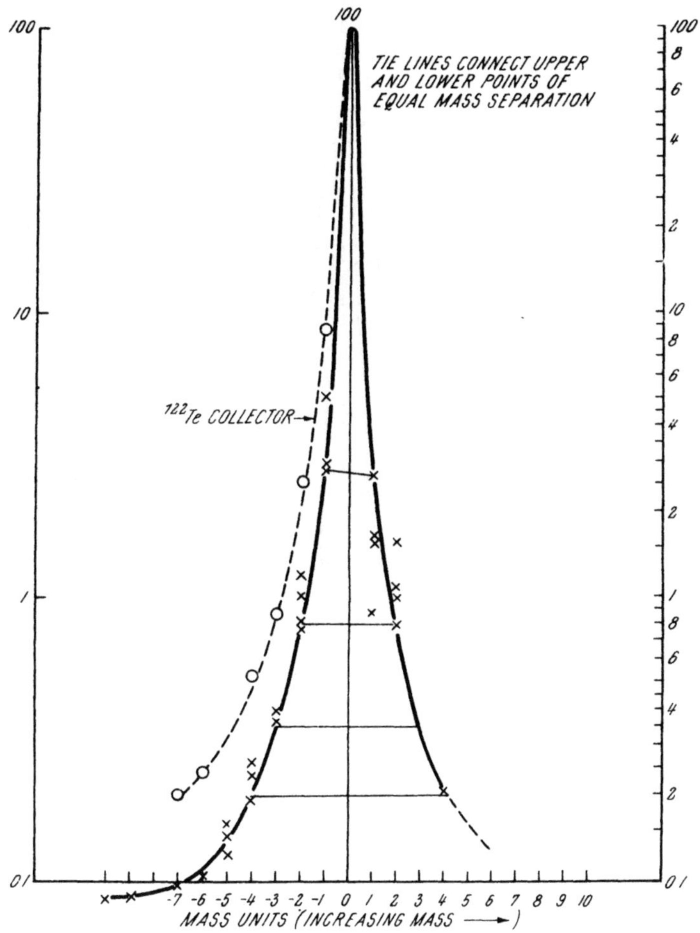

Fig. 10. Tellurium distribution (A. E. R. E.)

less than 0.1%. The significance of the Te^{122} curve has already been discussed (p. 51).

Fig. 11. The tendency of chromium to cause electrical breakdowns is demonstrated on the curve; about half of the lower mass contamina-

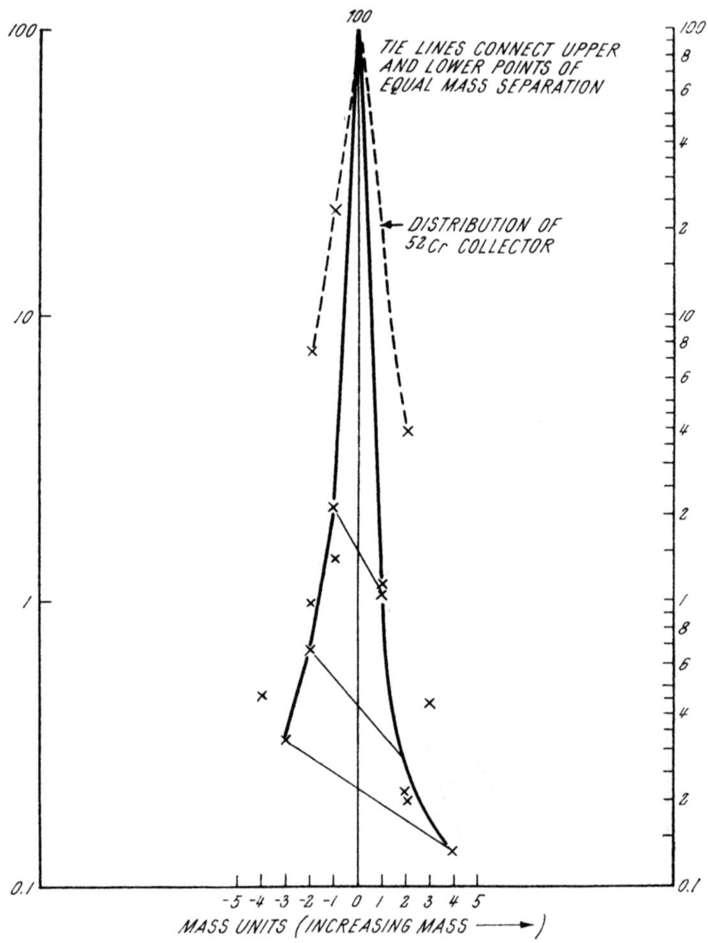

Fig. 11. Chromium distribution (O.R.N.L.)

tion being apparently due to this case. Source material contamination less than 0.1%. The significance of the Cr^{52} curve has already been discussed (p. 51).

Fig. 12. Shows the distribution curves, for the Ca^{46} collectors only, of two calcium separations carried out at A.E.R.E. The full line

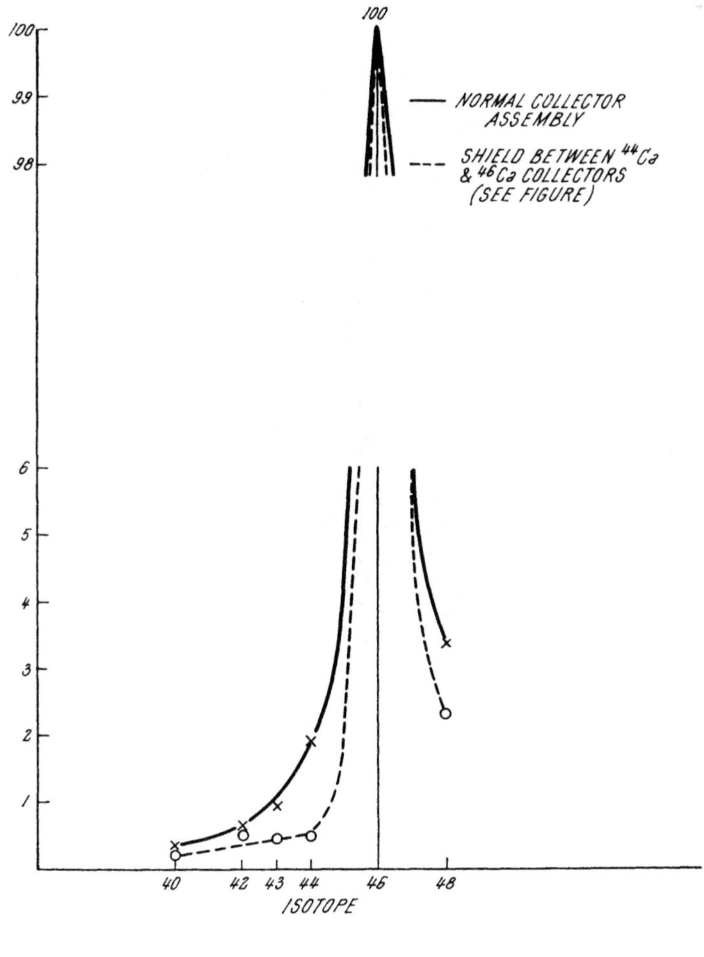

Fig. 12. Calcium distribution (A.E.R.E.)

represents a routine separation. The broken line was for a special collector assembly fitted with a shield (Fig. 13), to minimise the contamination of the valuable Ca^{46} by neutrals due to gas scattering [section (f)]. As can be seen, the curve in this case is much sharper on the shield side, and it is apparent that the shield has reduced the contamination by the abundant lower mass isotopes.

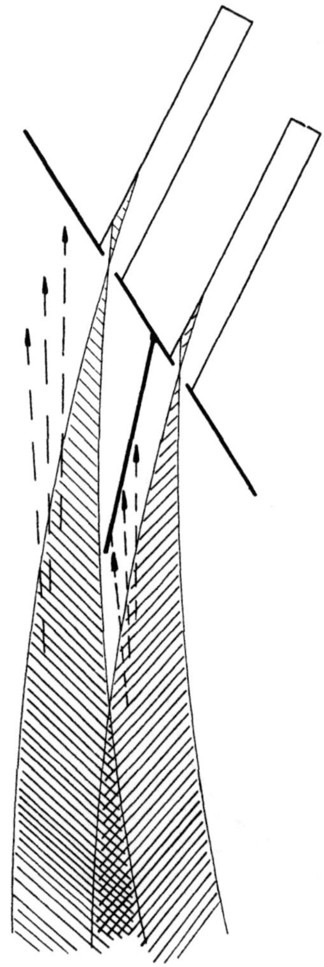

Fig. 13. Calcium collector with shield

Fig. 14. Shows three curves for plutonium separations on Hermes, the high resolution sector machine at A.E.R.E. As can be seen, sparking has in each case contributed considerably to the loss of resolution. Contamination by source material is very low (for h. 30 less than 0.1% of the total collection).

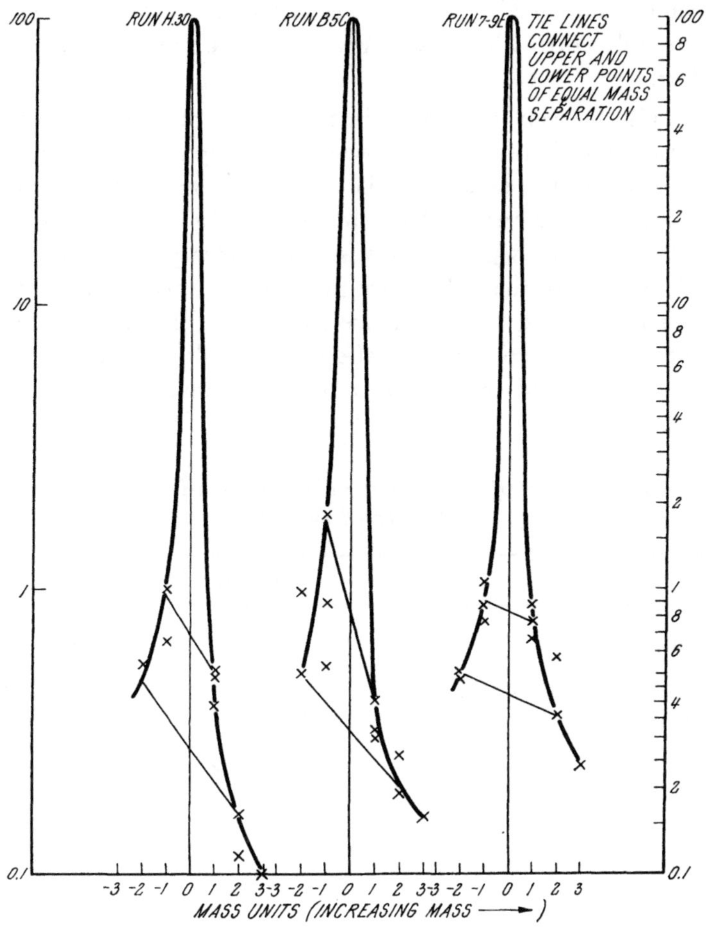

Fig. 14. Plutonium distribution (Hermes)

Discussion of Results

It is apparent from the curves that even without any data on the ion beam distribution, quite valuable information can be obtained. It can be seen that in a number of cases sparking appears to be causing a major contribution to the beam spread. If only the higher mass isotopes were required in a pure state (e.g. Ca^{46}, Pu^{241}, Pu^{242}, U^{238}), this effect would not be serious, but in normal separations, where this is not the case, the sparking would very seriously limit the ultimate enhancement obtainable. In high resolution machines, it therefore would be well worth while taking every possible precaution to reduce this effect.

It can also be seen that contamination by source material, whether by direct evaporation from the source unit, or by subsequent sputtering from the vacuum tank walls, is very low in most cases (although it is important with very volatile materials: Hg, Cl, Br, S etc.). It follows that in most separations it is not necessary to take any extra precautions against this effect. In the case of very high resolution separations, it would become more important. Table 7 below demonstrated the limitation of this effect on the ultimate purity of the sample, assuming all other causes of contamination to have been eliminated.

Table 7. *Limiting Effect of Source Material Contamination on Ultimate Purity of Separated Isotopes*

	Maximum Possible Purity					
Natural abundance	99%	90%	50%	10%	1%	0.1%
1% total collection due to source material	99.99	99.9	99	91	50	9
0.1% ditto	99.999	99.99	99.9	99	91	50

It can be seen from the Table that very low abundance isotopes can be well enriched if this effect is low (<0.1% of the total collection).

Conclusions

The relative enhancement coefficient has been shown to provide the basis of a valuable method for studying separator performance. The calculated beam spread can be used to predict or detect errors in mass analysis results. It appears that under normal conditions all of the isotope beams of a given separation are equally resolved, regardless of their relative intensities.

The calculated beam spreads can also be used to obtain an understanding of the causes of contamination which predominate in a separation. In particular it appears that in a number of cases electrical breakdown is one of the principal limitations of enhancement and that in general the contamination caused by source material is of minor importance compared to that originating from other causes.

Acknowledgements

The author would like to express his appreciation to LEON LOVE of Isotopes Separation Department, Oak Ridge National Laboratory, who kindly supplied many of the detailed analyses on which this study is based and also to Mr. D. F. DANCE of Mass Spectrometry Group A.E.R.E. Harwell, who carried out the mass analyses of U.K. Samples and also to R. W. MCILROY for many stimulating discussions and for helpful suggestions in the preparation of this report.

References

1. BERNAS, private communication.
2. LIVINGSTON, T.N.C.C. communication.
3. Principles of Chemical Engineering, p. 521. McGraw Hill.
4. BOYD WEAVER, U.S.A.E.C. Report, O.R.N.L. 1627.
5. LOVE, private communication.
6. KETRON, BELL and LOVE, U.S.A.E.C. Report, A.E.C.D. 3642.
7. DUNCAN and COBB, U.S.A.E.C. Report, Y-538.
8. LOVE and LEYSHON, U.S.A.E.C. Report, A.E.C.D. 3501.
9. LOVE, BELL and BUTTRAM, U.S.A.E.C. Report, Y-693.

Contribution to the Study of Isotopic Contamination with a Double Magnetic Deflection Isotope Separator

By

R. Bernas, J. Camplan, M. van Ments, and **J. L. Sarrouy**

Université de Paris, Laboratoire de Physique Nucléaire, Orsay (S. et O.)

With 8 Figures

Summary

The analysis of the isotopic composition of the ion beam which enters the collector of an electromagnetic separator gives an accurate indication of the nature of contamination in these instruments. We used for this study the second magnetic stage of our isotope separator.

Although the contribution of fast neutral particles produced by change of charge is important, it can easily be eliminated and does not constitute a limiting factor to isotopic purity. This limit is due to scattered ions and, for a small part, to ions that were slowed down or that did not receive the full accelerating voltage.

The contribution of neutrals is obtained from the difference between the isotopic composition of a deposit made on the first stage collector and that of the corresponding ion beam analysed by the second stage.

The experiments were carried out on Sr, Yb, and U. The results confirm the preferential contamination by neighbouring isotopes of heavier mass. The origin of this asymmetry is discussed. The consequences of these various processes on the limiting performances of magnetic cascades are studied.

Introduction

The study of the origin of isotopic contamination in electromagnetic isotope separators is necessary to establish projects of new instruments capable of very high performances and in order to improve the operation of existing machines.

One knows that the principal causes of that contamination are, besides the ion optics defects, the interaction between accelerated ions and the residual gas as well as space charge effects.

An important work in this field has been published recently [1] and it seemed interesting to us to study again some points more in detail and with the help of the double magnetic deflection isotope separator which we have recently built [2].

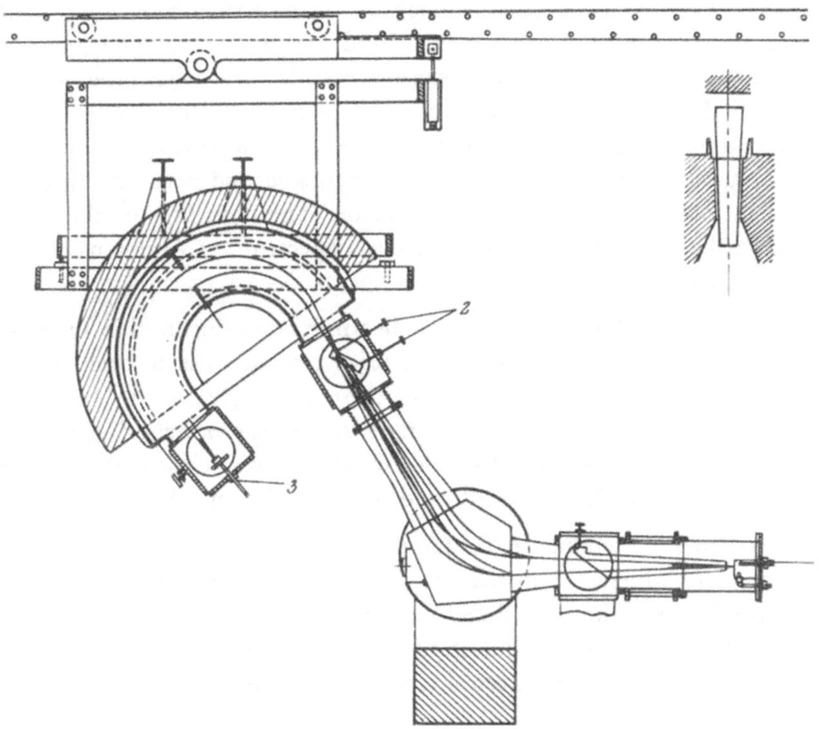

Fig. 1. Schematic drawing of the experimental set up

1 Ion source 2 Intermediate diaphragm 3 Final collector

Our study was carried out on the following points:

I. Relative contribution to the isotopic contamination of scattered ions and of fast neutrals.

II. Asymmetry of the contamination.

III. Causes of limitation to isotopic purity in the second magnetic stage.

IV. Experimental evidence of a process of contamination by the neighbouring isotope of heavier mass.

I. Relative Contribution of Scattered Ions and Fast Neutrals

(a) *Experimental set up*

Fig. 1 shows a schematic diagram of the instrument used. The ions are issued from the source 1 and analysed by a homogeneous magnetic field, then they are focused on collector 2 where the desired isotopic beam passes through a slit of 1 to 2 mm width. This beam can then either be collected on a metal surface, the deposit being analysed in a mass spectrometer, or penetrate into a second magnetic analyser which focuses it on the final collector. The scanning of this field will make it possible to take a mass spectrum of the ions which make up this beam.

In order to eliminate any mass discrimination during this analysis, it is necessary that the ion trajectories remain close to each other; this is obtained by limiting the aperture angle of the beam to approximatively 5° and by ensuring through appropriate diaphragms that the incidence of ions of various masses remains close to the normal (an exception to this case will be discussed in point 4).

In order to eliminate contaminations that could occur through chemical treatment and in order to simplify the experimental procedure we used a technique of direct deposition of ions on mass-spectrometer filaments (Fig. 2). These filaments are 1 mm wide and 10 mm long, and are then introduced directly into the analytical spectrometer (thermionic emission).

We were able to verify that, in the case of uranium, in particular, the composition of the deposit made on a classical collector was not different from that indicated by the thermionic filaments that were used during the same collection (Fig. 2b).

The adjustment of the filaments or crucible can be done without stopping the operation of the separator and it is always preceded or immediately followed by several analyses of the ion beam in the second magnetic stage.

The deposit at the first stage is due to ions of the desired mass, ions of neighbouring masses that were scattered, and to the impact of fast neutral particles created by change of charge in the space between the magnetic analyser and the collector. We shall neglect the contribution of neutral vapors issued from the source.

The result of the isotopic analysis of this deposit, compared to the mass spectrum of the beam obtained at the second magnetic stage,

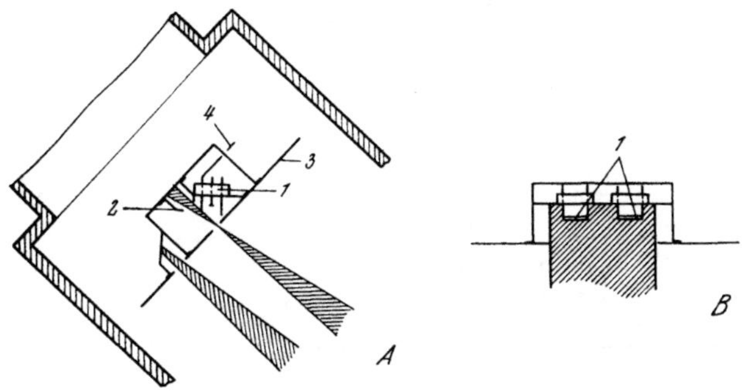

Fig. 2. A device for collecting the ions on a mass spectrometer filament

1 Triple filament adapted to the mass spectrometer ion source (the current is measured through connection 4)

2 Small collector pocket for direct deposition of the beam not collected on 1

3 Intermediate diaphragm of the separator

B The respective positions of the two triple filament assemblies and of the collector pocket

makes it possible to distinguish, *in a first approximation*, the part of contamination due to scattered ions and to neutrals. *The approximation results from the fact that slowed down ions in the first stage may have the same momentum as the ions of the main beam and they will experience in the second magnet an identical deflection, so that their number cannot be evaluated in these experiments.*

Our study was carried out on Sr, Yb and U.

Ion intensity at the first stage collector varied between 0.5 and 1 mA for the sum of the isotopes of the elements and the pressure in that stage of the separator was 1 to $2 \cdot 10^{-5}$ mm Hg. The results obtained are shown in Table 1.

Table 1

Strontium	84	86	87	88	Enrichment factor* at γ 1st stage	γ' after elimination of neutrals	$R = \frac{\gamma'}{\gamma}$
natural	0.56	9.86	7.02	82.56			
Sr^{87} deposit at 1st stage	—	—	95.8	4.2	250		3.3
Mass spectrum at 2nd stage	—	—	98.6	1.4		830	
	neutral contamination 66%						
	ionic contamination 33%						
Ytterbium	170	171	172	173			
natural	3.03	14.31	21.82	16.13			
Yb^{170} deposit at 1st stage	95	5	—	—	90		4.0
Mass spectrum at 2nd stage	98.7	1.3	—	—		360	
	neutral contamination 79%						
	ionic contamination 21%						
Uranium	235		238				
natural	0.7		99.3				
U^{235} 1st stage	62		38		220		3.1
Mass spectrum at 2nd stage	85		15			680	
	neutral contamination 72%						
	ionic contamination 20%						

* Defined as the ratio of isotopic composition before and after separation.

These results tend to show that contamination by neutrals is predominant in this type of instrument. However, the ionic contamination is of such a magnitude so that the elimination of neutrals will not increase greatly the enrichment factors.

The values we have obtained in these experiments are quite different from those published by CASSIGNOL [1] who obtained a ratio of $\gamma'/\gamma=25$.

We shall note that the value $\gamma'=680$ which we obtain is close to that obtained by the mixed cascade of Saclay ($\gamma'=500$), so that it must be in an unsatisfactory performance of the first stage of that instrument that one must find the cause of the discrepancy between these two values. In fact we operated with the same ion beam in the first stage (0.3 mA) and during a similar period (1 to 2 h) as that indicated by this author.

A typical mass spectrum obtained during this study is shown in Fig. 3.

A new study is being carried out to determine the effect of an increase in pressure on the ratio of neutrals to ions and on the part played by slowed-down ions [3].

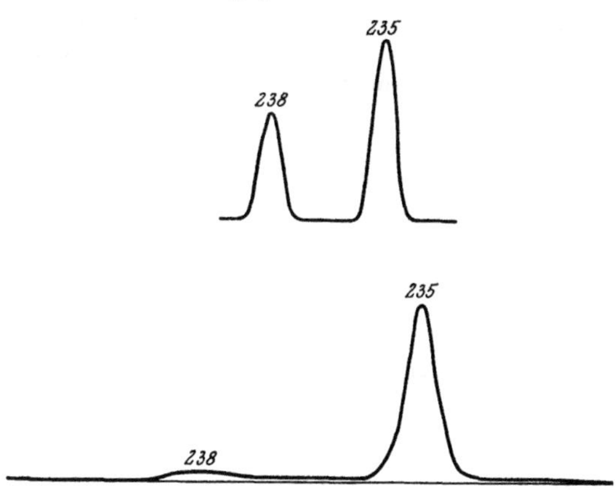

Fig. 3. Mass spectrum of an uranium ion beam

above: Mass spectrum of a U^{235} *deposit* on a filament at first stage
below: Mass spectrum of the same beam at the collector of the 2nd stage

II. Asymmetry of Contamination

The analysis of spectra obtained with the second magnetic stage confirms the predominance of the heavy component in ionic contamination. We reach the same conclusion also as regards contamination by neutrals.

Table 2 below corresponds to the analysis of an ion beam of Yb^{173} entering into the intermediate diaphragm which shows that the enrichment factor of Yb^{173}, with respect to Yb^{172}, is in both cases (ionic or neutral contamination) superior to the enrichment factor determined with respect to Yb^{174}.

Table 2. *Isotopic composition of the Yb^{173} ion beam at the collector at the first stage*

Masses	A after elimination of neutrals		B after elimination of scattered ions	
172	0.3		1.3	
173	99.7	99.4	98.7	96.4
174		0.6		3.6
Corresponding enrichment coefficient	460	320	112	50

* For the determination of the enrichment factors we consider only a pair of isotopes at a time: 172 and 173 or 173 and 174 neglecting the presence of the third one so that the indicated proportions in column A and B are expressed in % for each pair of the isotopes.

An indication of the origin of this asymmetry is given by the following experiment:

An ion beam of Yb^{173} with its impurities enters the second stage and its spectrum at the final collector is given in Table 3.

Table 3

Mass number	172	173	174	
Abundances in natural Yb (%)	22	16	32	
Composition of the ion beam Yb^{173}	2.2	825	4.5	$p = 1.5 \cdot 10^{-5}$ mm Hg
Ditto	45	930	95	$p = 8 \cdot 10^{-5}$ mm Hg

When the pressure in the instrument passes from $1.5 \cdot 10^{-5}$ mm Hg to $8 \cdot 10^{-5}$ mm Hg the ratio Yb^{172}/Yb^{174} is constant. It seems then that the asymmetry of contamination originates from a process of interaction between the ions and the residual gas and is not due to hash or to incomplete neutralization of space charge.

One must note that of all elements that were studied (Cr, Sr, Eu, Yb, Hg, Pb, U) only lead shows a preferential contamination by the neighbouring light isotope (Fig. 4) and this did not come from an error in centering the beams in the first stage diaphragm. The origin of this anomaly is not yet known.

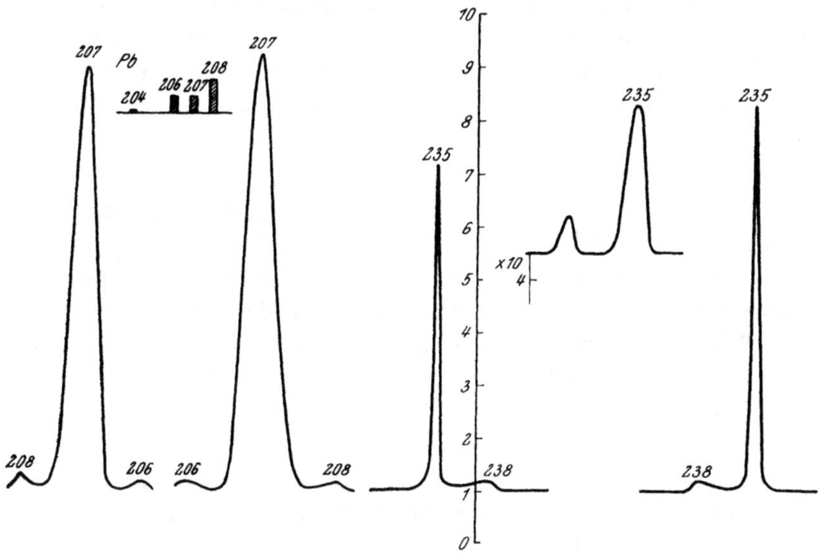

Fig. 4 Fig. 5

Fig. 4. Mass spectrum at the second stage of a lead ion beam. This spectrum shows a definite preferential contamination by the neighbouring isotope of lighter mass
above: Natural isotopic abundances of lead

Fig. 5. Mass spectra of U^{235} obtained under the same conditions as those of Fig. 3 but after the introduction of a diaphragm limiting the angular aperture of the beam to 1.5°
above: Mass spectrum of *deposit* on filament
below: Mass spectrum at the second stage
Notice the tail of the mass 238 peak towards 235

We analysed the isotopic composition of the deposit obtained at the collector of the second stage by collecting the ions on very carefully purified mass spectrometer filaments. The results of the analyses show a definite preferential contamination by heavier neighbours. Though we should not be surprised by this result, since ionic contamination at the entrance to the second stage is in favour of heavy isotopes, the examination of the U^{235} peak reveals, however, a supplementary reason for this preference: the considerable spread of U^{238} towards U^{235} (Fig. 5).

III. Study of the Causes of Contamination After the Double Deflection

In order to ascertain that the spread of the beams just mentioned would not be due to a radial aperture aberration and in order to eliminate any abnormal incidence of the ions of U^{235} we introduced a new set of 3 diaphragms at the intermediate focus in order to limit the radial aperture angle to 1.5°. The new mass spectra obtained are shown in Fig. 5.

The spread of the 238 peak is again very clearly visible and we are then led to consider the existence of an inhomogeneity in the energy of the ions of U^{238}. The process which is at the origin of this inhomogeneity is not yet known. None of the hypotheses proposed so far seems valid. Considering the small value of the angle of incidence of U^{238} ions imposed by the intermediate diaphragm of the first stage, the energy loss should occur before or in the first analyser. In the space between the analyser and the collector, one would have to rely on a multiple scattering process.

The isotopic analysis of U^{235} collected at the second stage indicates the presence of an amount of U^{238} varying between 1 and 2%. The contamination then is due to about 1/10,000 of the ions of U^{238} which reach the collector of the first stage.

This leads us to look among the possible causes of contamination for an interaction between ions and molecules which one may have neglected so far. It then seems absolutely necessary to study more profoundly the processes that can be at the origin of isotopic contamination before one can calculate the limiting performances of magnetic cascades. Such a study is at present under way [3].

The experiments just described point out clearly that it is not possible simply to multiply the enrichment factors of each of the stages together.

IV. Experimental Evidence of a Process of Contamination by the Neighbouring Isotope of Heavier Mass

Fig. 6 represents the position of the ionic beams in the neighbourhood of the intermediate diaphragm during the experiments we describe.

The study[1] was carried out on cadmium beams; Cd^{113} entered in the second stage and a mass spectrum of the charged components

[1] This study was carried out in collaboration with M. Cassignol [5].

of this beam was recorded. The mass spectrum obtained under normal operating conditions ($p = 1 \cdot 10^{-5}$ mm Hg, $i^{+}{}_{\mathrm{Cd}} = 1$ mA) is

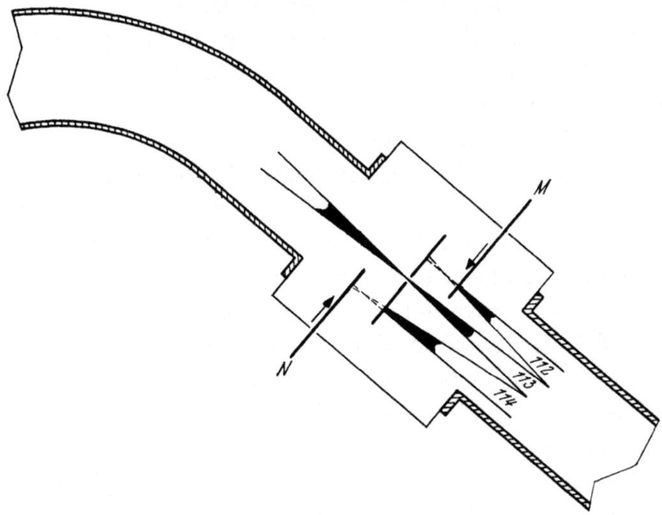

Fig. 6

shown in Fig. 7. The mass 113 peak is accompanied by a low-intensity satellite of mass 114 and we were able to check that the satellite did, in fact, belong to mass 114. The spread of the 113 peak towards

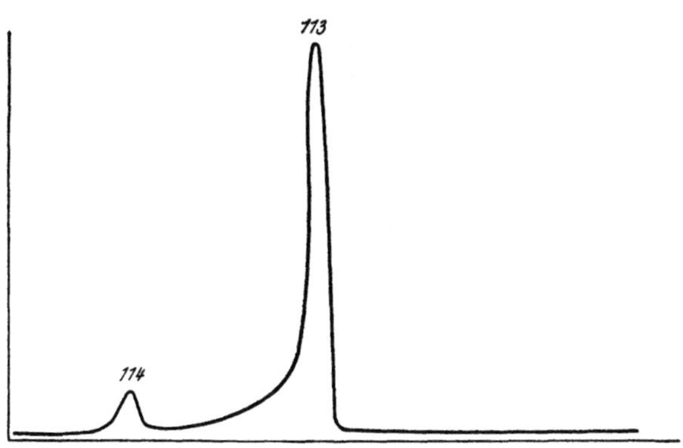

Fig. 7

the 114 peak was primarily due to an aperture aberration (the second stage was not corrected for this aberration during this experiment).

(We shall see later the importance of this state of matters in the experiment which we describe.)

When the pressure in the first stage was increased to $5 \cdot 10^{-5}$ mm Hg the intensity of peaks 113 and 114 decreased and 2 new peaks appeared, one on each side of the 113 peak (Fig. 8).

The mass identification definitely showed the two peaks to be due to masses 114 and 112 and to be caused by scattering without change of charge in the first stage.

However, the origin of the first peak of mass 114 is different: its position at a greater distance from mass 113 than that corresponding to the peak of the scattered mass 114 implies that it is the result of a parasitic beam of mass 114 which enters into the second analyser with an oblique incidence *and from the light-mass side.* We can then see that this beam will travel for at least part of its trajectory in a region where the field is decreased, due to edge losses, so that its impact on the collector will occur beyond those ions of mass 114 that remained in the unperturbed field region.

We checked this hypothesis by introducing metallic plates close to the mass 113 beam before and after the intermediate diaphragm (Fig. 6).

If one moves plate *M* radially from the outside towards the intense part of the 113 beam we observe the following effects on the mass spectrum of Fig. 8. The peak of mass 114 which is furthest away

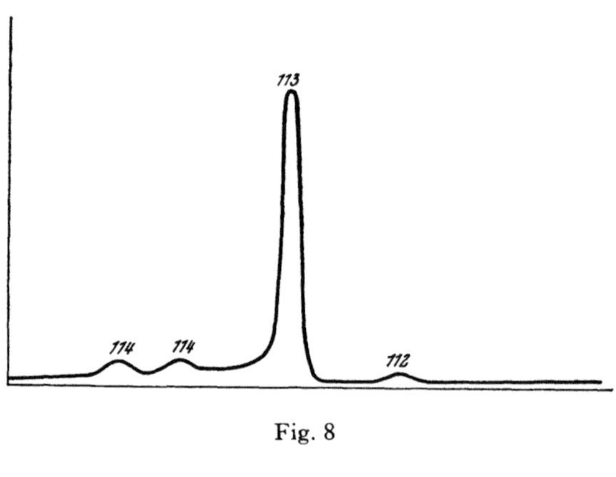

Fig. 8

from peak 113 vanishes first without any attenuation of the other peaks. If one penetrates further into the beam, peaks 112, 113, and 114 decrease progressively and simultaneously[1].

[1] The phenomena are identical if plate *N* on the other side of the diaphragm is brought closer to the beam.

This observation is of great interest for the improvement of the performance of isotope separators. The introduction of a diaphragm after the collecting slit can permit a very considerable increase of enrichment factors without the need of a second stage. However, it is clear that the double magnetic deflection separator used in these experiments eliminates these defects even in the absence of an aperture diaphragm. Performances of electrostatic-magnetic cascades can be improved through the use of the above method. This has been confirmed recently by some experiments on mercury [4].

A hypothesis on the origin of this phenomenon has been proposed [1].

Note added in proof:

The continuation of this work [3] has shown that a considerable part of the contamination, contrary to what was accepted up to now, is due to low-energy ions. This implies that the values given in this paper concerning the relative contribution of neutrals and ions to the contamination must be revised.

References

1. Cassignol, C., C.E.A. report No. 1185 Saclay (1959).
2. Bernas, R., et al. see: Double Magnetic Deflexion Isotope Separator for the Production of Very High Purity Isotopes.
3. Bernas, R., J. Camplan, M. van Ments, J. Phys. Radium (to be printed) 1961.
4. Cassignol, C. private communication.
5. Bernas, R., and C. Cassignol, J. Phys. Radium **21,** 566 (1960).

Discussion

M. J. Higatsberger: The elegant method of collecting isotopes on a filament and then analyzing the deposit of the filament requires a very homogeneous beam, otherwise one could easily get results which are different from the collected material in the pocket. Furthermore, I should like to recall that thermal ionisation is dependent on many factors. Under ideal conditions the ratio of evaporated isotopes is proportional to the square root of their masses.

R. Bernas: In the particular case of uranium we collected simultaneously on a collector and on a filament. We analysed both deposits and found no difference between the two. Of course, one has to be careful to set the filament in the right position. We are aware of the effect of mass discrimination in thermionic emission. We therefore use for this method only heavy elements as e.g. Yb or U where the square root of the mass effect is negligible.

L. O. Love: What was the relative size of the filament to the target?

R. Bernas: The slit has the dimensions of 1.0–1.5 mm by 40 mm. The filament was 1 mm wide and 10 mm long. We usually put 2 filaments into the slit and leave the centre part for direct collection. In the case of contamination by source material, the relative areas of the filament and of the collector may have great effect. Therefore, we made an analysis of a collected sample of uranium and at the same time an analysis under the conditions of the deposit at the filament. The agreement was better than 1%.

L. O. Love: What was the angle of acceptance from the beam of the source?

R. Bernas: About 2.5°.

P. Weinzierl: All the preceding papers seemed to show the importance of the neutral component of the beam in isotope contamination. To remove this component is one advantage of two stage machines. Would it not be possible to separate ions and neutrals at least in a rough way inside a collector? If only one isotope is needed in the separation, there should be space enough available to provide something like a retarding and deflecting field within the collector. This might replace, to some extent, a second stage separator.

R. Bernas: We have done such experiments but with a very crude system. By using a small condenser just after the slit of the collector and deflecting the beam by a few degrees, we only got an increase in the enrichment factor from 320 to 360 in the case of Sr. So we stopped these experiments but we may do them over again more properly.

J. H. Freeman: We are trying to reduce neutral contamination by utilizing the bending effect of the calutron magnet. The beam is deflected inside the collector and collected in a different place to the undeflected neutrals.

Ch. Cassignol: There is no agreement between Dr. Bernas and myself about the question of energy spread. Would anybody care to comment on this subject, especially to what extent the energy spread may be responsible for isotopic contamination?

M. J. Higatsberger: Has anybody tried to analyse the energy spread with an electrostatic analyser?

Ch. Cassignol: When comparing the contamination by heavier isotopes in ordinary separations with the results of our direct measurements of energy spread using such an electrostatic analyser, it was found that the latter could not by several orders of magnitude explain the former. An alternative explanation had to be found, and I tried in my paper to describe a possible mechanism. Experiments of the Orsay group indicate that the accuracy of energy spread measurements may not have been sufficient. It is planned to repeat direct measurements with a more accurate and collimated electrostatic optics and with an increased sensitivity for the current measurements. (See note added in proof p. 32.)

Discussion about the topics I 3 — I 6

R. W. McIlroy: There was a discussion about enrichment factors. I would suggest that there are two applications for the enrichment factors. One to describe what the machine has already done and one to describe what the machine may do in future. The conventional approach is to speak of an enrichment factor simply in terms of concentrations, this is what the customer wants and this is what one wants to tell him. He wants to know how much of the required isotope is in the mixture and how much of the unrequired one. Whereas for descriptions of what machines may do in future Dr. Nielsen's suggestions seem useful and they relate possible enrichment factors to machine parameters. I would, therefore, suggest that when we speak of enrichment factors we speak of concentration only and also that we use relative concentrations as between two components only; Dr. Nielsen's type of work is essentially an evaluation of enrichment factors in terms of machine parameters.

R. Bernas: We had left a question open about the enrichment factor given by Dr. Alväger. We had two discussions together and finally it turned out that his enrichment factor did take into account the relative abundance of the neighbouring isotopes. However, there are some assumptions made on the shape of the peaks of the main and of the small isotope. I came to the same conclusion as Dr. McIlroy: I wish that we could make use of the fact that we are all gathered here to define in common the "enrichment factor" which should be used in the future. That would very much simplify the matter. So I think we should only make use of concentrations of isotopes that have been effectively separated. Whether these measured concentrations were determined by radioactive measurements or by mass spectrometry does not matter. Then I think we could also try to specify two different durations of collections when talking about enrichment factors so that one should not compare the enrichment factor of a 24 hour collection with that of a 1 minute collection. We could perhaps agree on two values: 1 hour in one case and just 10 minutes in the other case. That would make it easier to compare instrument performances in the future.

J. H. Freeman: I think we can probably agree to use relative enrichment factors by whatever symbols we normally use as long as it is understood that we are talking about the relative factor and not the normally defined factor. I don't know whether we should determine a certain collection time since the operating conditions vary very much. One could at least ask that the separation time be given with the enrichment factor.

K. O. Nielsen: I suppose you may also say something, if you use relative enrichment factors, about the difference between the two masses. If you want to introduce the separation time why not also introduce the capacity? If you go down with the current much below the normal current of the isotope separator you can increase the enrichment factor.

R. W. McIlroy: I think the enrichment factor should be defined as the ratio of concentrations without any restrictions.

L. O. Love: What we are considering here is establishing a method for measuring isotopic enrichment factors that might be very difficult to change, if inadequate, in years to come. It seems to me advisable to have a committee consider this problem and report to this group their recommendations.

W. Walcher: I think the discussion on the enrichment factor is fruitless unless this term means exactly the same to each speaker. Every physical quantity is defined by an equation. At first we should write down the defining equations of all the quantities suitable for characterizing the result of the separation process and subsequently we should introduce terms for these quantities. In doing that we should alter the meaning of any term already in use.

J. Koch: I think the idea of Dr. Love is in a way good but since we are only a small group working in this field we might postpone the question for some time and rather try to get together soon again in a new symposium, where among others these questions may be settled.

J. Kistemaker: I can hardly see how Dr. McIlroy can be misunderstood in his very simple and succinct statement. There is only one definition in all isotope separations not only in an electromagnetic separation but in all the other ones. It is a simple definition of a process X which happens, namely the ratio of the concentration of two isotopes before the process and after the process. I hardly can see how there can be any discussion on this subject.

Chapter II

Electromagnetic Separators Planned

Electromagnetic Separator for Radioisotope Investigations

By

N. J. Freeman, W. A. P. Young, R. W. D. Hardy, and **H. W. Wilson**

U. K. A. E. A., Aldermaston

With 2 Figures

Summary

Details are given of the vacuum system, magnet, ion source, electrostatic focusing system and electrical supplies of a 100 cm radius, 90° sector, electromagnetic separator which is now under construction at A.W.R.E., Aldermaston. The machine is designed to produce milligramme quantities of stable isotopes and radioactive isotopes up to 1 millicurie strength. An effort is being made to secure a higher vacuum than has been used in the past, with the principal objects of obtaining a higher resolution and enrichment factor.

Introduction and Specification

The designs of recent electromagnetic isotope separators have been influenced largely by the desire for purity rather than quantity of the collected samples. The development of the low intensity isotope separators by the Scandinavian groups is a good example of the progress which has been made in this field. Since these machines have many features which are desirable for the production of active and stable isotopes suitable for use in nuclear physics experiments, the design of the A.W.R.E. separator has followed these instruments in many respects.

The principal requirements were: (a) the separation of radioactive material into samples for use in decay scheme studies or as radioactive tracers, (b) the production of stable isotopes for nuclear reaction measurements. The machine has, therefore, been designed to handle activities up to about 10 mC and collection rates for stable isotopes will be a few milligrammes per day. In addition to the simultaneous collection of all stable isotopes of a given element, a high overall efficiency (~10%) of separation of small samples was required to reduce the collection times and the total activities to be handled. A control on the height of the focused mass lines and a system for continuously monitoring the ion beams were also desirable.

High enrichment factors can be achieved by minimising (a) the aberrations of the electrostatic lenses, and the magnetic field and (b) the interactions of the ions with the residual gas in the vacuum system. The latter is particularly important, and we are attempting to achieve suitable beam shapes by using high acceleration voltages (up to 70 kV) and a high vacuum ($<10^{-7}$ mm Hg). These requirements combined with magnet construction costs decided that the mean beam radius should be 100 cm. This gives a dispersion of 5 mm for adjacent masses at mass 200, and a resolving power of about 1000 should be obtained.

Description of Machine

Magnet

The 100 cm radius 90° sector electromagnet was constructed by A.E.I. (Manchester) Ltd., and weighs about 8 tons. Fields of up to 6000 oersted are produced in the 6 cm gap by high impedance coils fed from an electronic series valve stabilizer supply producing 1.8 kV at 2.3 kW. The field is measured and stabilized to a few parts in 10^5 by the proton resonance method. The pole pieces are 30 cm wide and the uniformity of the field over a radial distance of ± 5 cm about the centre line is better than 0.03%.

The magnet is supported on hydraulic jacks for small vertical adjustments, and the complete assembly is mounted on wheels and can be moved a distance of 4 feet along a line parallel to the axis of the electrostatic lens system (see Fig. 1). This refinement will simplify outgassing of the ion path and dismantling if the machine contamination rise sabove tolerance. To find the optimum focusing conditions the magnet can be moved horizontally on its supporting surface table. The friction is reduced by forcing air at 25 p.s.i. between the sliding surfaces.

The magnet has been designed so that the pole pieces can be easily removed, and plans are being made to use shaped pole pieces for double focusing. In this case the collector will be 230 cm from the magnet boundary, and the dispersion will be increased by a factor 2.

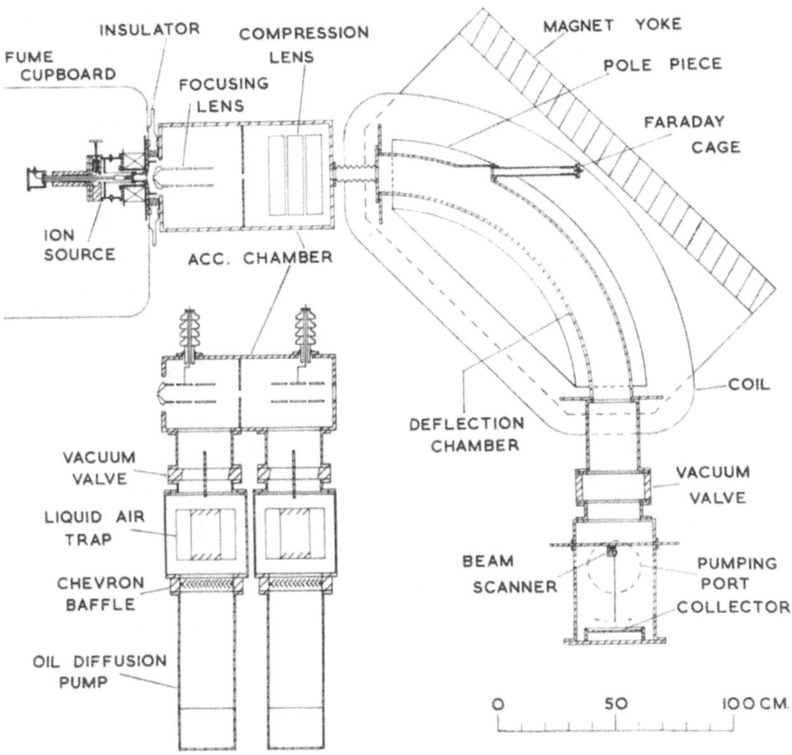

Fig. 1. Constructional Details of the Isotope Separator

Vacuum System

The vacuum system has been constructed almost entirely from welded stainless steel which has low outgassing properties and will be relatively easy to clean if the machine becomes badly contaminated. Lead and p.t.f.e. gaskets have been used for all static joints while rubber 0-rings provide seals for the large valves and the demountable joints on the source and collector.

To achieve the required low pressures, a high pumping speed is supplied by three 9-inch fractionating oil diffusion pumps, each with a speed of 1500 l/sec. A freon cooled chevron baffle and a high conductance liquid air trap are fitted above each pump to reduce the back streaming of oil vapours and to ensure a high pumping speed for condensible vapours in the system. Two of the pumps are connected to the acceleration chamber (Fig. 1) which has an adjustable aperture at the centre to give differential pumping. By this means, pressure rises due to the hot ion source are limited to the source region. Tests have shown that pressures less than 10^{-7} mm can be obtained without outgassing.

Ionization and Penning gauges measure the pressure in the main system while Pirani gauges are connected into the backing lines. Remotely operated magnetic valves control the pumping lines and the usual thermal protection devices and safety circuits are incorporated.

Ion Source and Power Supplies

The ion source assembly and sample loading mechanism has been modelled on the design of NIELSEN (1957). The source is mounted inside a fume cupboard at source potential (70 kV), and is insulated from the main vacuum chamber by a composite Mycalex and Araldite insulator.

The electrical supplies for the ion source are provided by a 5 kVA three phase isolating transformer. Three rectified low voltage supplies giving up to 75 A are available for the discharge tube filament and heater and for the furnace heater. The furnace temperature is measured and regulated automatically by the output of a chromel-alumel thermocouple. The arc supply (300 V 1.5 A) is voltage stabilized and the arc current may be controlled manually or automatically by a magnetic amplifier feedback loop in the filament supply. A 650 V 650 mA (D.C.) current stabilized supply is provided for a source magnet which can produce an axial magnetic field of up to 1000 oersted.

The source is cooled by circulating oil through copper pipes soldered to the source assembly and a water cooled heat exchanger at earth potential. Solid samples can be inserted through a vacuum lock which is evacuated by a 1-inch diffusion pump and a rotary pump at ground potential. Electrical insulation for both oil and vacuum lines is provided by the use of polythene tubes.

Source investigation work has been concentrated on a discharge chamber of the Unoplasmatron type (VON ARDENNE, 1956) which does not require a magnetic field. This discharge chamber (Fig. 2) is constructed from graphite, pyrophyllite, and molybdenum with a stainless steel exit plate. The filament is a helix of 1 mm diameter tantalum wire.

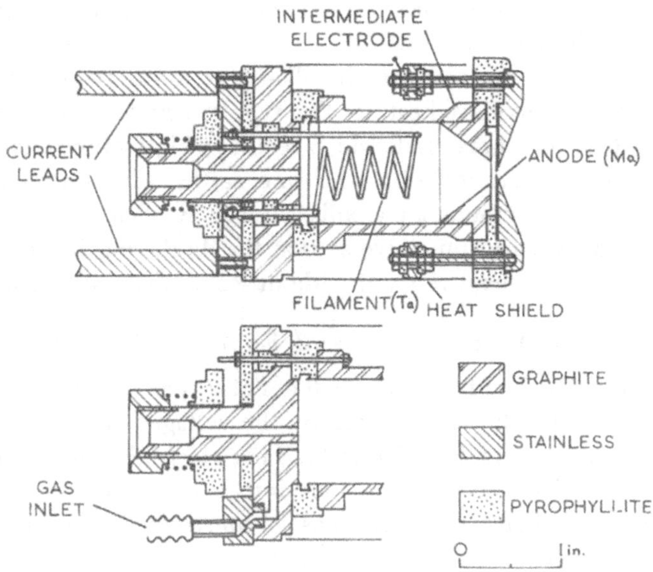

Fig. 2. The Unoplasmatron Discharge chamber

In this source, a low voltage arc discharge is constricted by the intermediate electrode which has an aperture of 5 mm diameter. The highly concentrated plasma on the anode side penetrates this opening and is separated from the rarefied plasma around the filament by an electrical double layer. Electrons from the rarefied plasma are accelerated through the double layer and are focused in front of the ion emission aperture where the intense plasma is formed.

The potential of the intermediate electrode is determined by a 500 ohm resistance connected to the anode supply. Typical operating conditions are: filament current 26–28 A; Arc 100 V 1 A (including approximately 150 mA to the intermediate electrode); minimum operating pressure $\sim 10^{-3}$ mm Hg. Currents of 800 μA can be extracted from a 1.5 mm diameter ion emission aperture.

The characteristics and the efficiency of the ion source and focusing system have been measured with argon gas and solid samples of lead chloride containing a Pb^{212} radioactive tracer. Efficiencies up to 20% have been obtained for argon and 9% for $PbCl_2$ with a discharge chamber of length 5 cm. The latter measurement includes both molecular and atomic ions since no magnetic analysis was performed to determine the relative concentration of these components in the ion beam.

Since the temperature of the discharge tube anode can reach 900° C without auxiliary heating, this type of source should be useful for the separation of materials which require a high source temperature to prevent condensation (e.g. the rare earths). In addition, as the source operates without an auxiliary magnetic field, it is relatively easy to arrange a vacuum valve so that the discharge chamber can be withdrawn for filament renewal without losing the main vacuum.

The Acceleration and Focusing System

Ions are extracted from the small circular aperture in the discharge chamber and are then accelerated by the voltage between the source and the earthed extraction electrode. This electrode forms part of a single lens which focuses the ions into a cylindrical beam with a divergence of about 0.5°. The lens is constructed of stainless steel tubes of diameter 7 cm and is mounted on a kinematic trolley so that the extraction distance is variable from 0.5 to 6 cm. Mycalex insulators are used to reduce outgassing, and the potential of the central electrode may be varied from +20 to +60 kV to optimise the resolution.

The height of the mass lines can be adjusted by the positive potential on the central electrodes of a compression lens formed by three pairs of stainless steel plates 30 cm long, 7 cm separation. Electrolytic tank experiments showed that this large ratio of length to separation was necessary to avoid distortions of the equipotentials at the centre of the lens by the proximity of the vacuum chamber walls. The outer pairs of plates are used to control the vertical position of the beams at the collector by connecting them to a ±500 volt supply.

The acceleration and lens voltages are supplied by a three phase selenium rectifier set producing 85 kV, 50 mA. The acceleration voltage is series stabilized with a stability of better than 1 in 10^4. In normal running conditions, however, the supply is automatically regulated by the signals from two pins placed at the collector on either side of a focused ion beam. The focusing and compression lens voltages are supplied by separate parallel stabilizers.

Collector Unit

Most of the work anticipated for this machine will involve the direct collection of the separated isotopes on thin foils. The collector plate is, therefore, a simple slide which can be moved to any desired position across the beams. If needed, masking apertures can be placed in front of the foils attached to the slide. A shutter is arranged to cut off the beams during the initial focusing operations. A small magnetic collector with an adjustable slit can be used to determine the resolution, or the focus can be observed by viewing a fluorescent screen attached to the shutter. In addition, the resolution and beam heights will be continuously monitored by a scanning device similar to that developed by NIELSEN and SKILBREID (1957). The reference pins which regulate the acceleration voltage are also mounted on a movable slide near the collector plane.

A valve in the collector arm is used to isolate the collector chamber from the main vacuum system so that samples can be removed immediately after separation. This is particularly desirable when short-lived activities are being investigated.

Acknowledgements

We wish to express our thanks to Messrs. D. E. P. SILVER, R. HAYES, A. D. BROWN, and other members of the electronics section for their contributions to the design of the electrical supplies and safety circuits. We also acknowledge the collaboration of W. H. HILLS and N. I. BROOME of the engineering section in the mechanical design of the machine.

This paper is published by permission of the Director of the Atomic Weapons Research Establishment, A.W.R.E., Aldermaston.

References

1. NIELSEN, K. O., Nuc. Instrum. **1**, 302 (1957).
2. NIELSEN, K. O., and O. SKILBREID, Nuc. Instrum. **1**, 159 (1957).
3. VON ARDENNE, M., Tabellen der Elektronenphysik, Ionenphysik und Übermikroskopie, p. 543, Berlin, Deutscher Verlag der Wissenschaften (1956).

Discussion

O. SKILBREID: Have you any experiences using pyrophyllite as an insulating material in ion sources? We use practically the same ion source as you and found that at temperatures of the order of 1000° C we got a lot of impurities from pyrophyllite especially when using chlorides and chlorine.

With pure silicon dioxide as insulating material one does not get any impurities at all at similar temperature conditions to those mentioned above.

N. J. FREEMAN: Since a magnetic analysis of the ion beams has not been performed we do not know whether impurities from the pyrophyllite appear in the ion beam.

J. KOCH: I should like to know whether the close proximity of the pumps to the acceleration chamber would cause discharges, as earlier experienced in the Stockholm separator. Or do the designers believe that the liquid air trap takes care of this possibility?

J. UHLER: The troubles at the Stockholm separator were mainly due to the old ion source. We do not believe that they came from the pumps. Nowadays we have no such trouble.

K. O. NIELSEN: You mentioned that the efficiency of the ion source does not depend on the length of the discharge chamber. Do you mean the length of the whole discharge chamber or only the small space between the intermediate electrode and the anode?

N. J. FREEMAN: The ion source is of preliminary design. No changes in the distance between the intermediate electrode and the anode have been tried, but increasing the total length of the discharge chamber tends to decrease the efficiency. When the separator is completed a systematic investigation of the ion source characteristics will be made.

K. O. NIELSEN: What form has the magnetic field in the unoplasmatron ion source?

N. J. FREEMAN: There is no magnetic field in the ion source.

R. BERNAS: What provisions have you made for decontamination of the chamber of your separator? You are dealing with hundreds of millicuries — that is very much.

P. WEINZIERL: In the case of the Vienna separator the idea on the amount of β and γ activities in the separator is about as follows: We want to start first of course, with the machine running on stable isotopes. After this preliminary use we plan to go over to rather small activities. The activity to be separated for nuclear spectroscopy work is usually of the order of 1 mC or less. Working with reactor-irradiated samples, frequently the induced activity will be one or two orders of magnitude higher than the source activity required. We should like to gain practical experience on the use of liners in the vacuum chamber and how to get out the activity from the chambers and the pumps. That's the reason why we are starting with small activities. After the preliminary studies we should be able to handle activities as high as 100 mC.

N. J. FREEMAN: A similar answer will be true for our machine. Initially, stable isotopes will be used and later a maximum β and γ activity of 10 mC will be handled in normal operation. The fume cupboard around the source will minimize the risks of atmospheric contamination when the discharge chamber is dismantled. A second fume cupboard may be constructed around the collector unit.

The Design of the Electromagnetic Isotope Separator of the S. G. A. E.

By

F. P. Viehböck

Österreichische Studiengesellschaft für Atomenergie Ges. m. b. H., Vienna

With 7 Figures

A. Introduction

The E.M. Isotope Separator of the S.G.A.E. will be a medium size machine with a total ion current at the target of about 5 mA for medium masses. The total radioactivity to be handled in our machine at one time will be of the magnitude of several hundred millicuries.

When separating radioactive isotopes, it is very desirable to have a large enrichment factor as well as a high luminosity of the machine. Because of the greater mass dispersion and the two-directional focusing property we use an inhomogeneous magnetic field for a mass separating field [1]. When working with radioactive isotopes, it is advantageous for handling purposes to have the ion source and the collector outside the magnetic field. Therefore, we decided to use a magnetic sector field. The ion optical properties of inhomogeneous magnetic sector fields, including the trajectories out of the medium plane, were calculated by Tasman, Boerboom, and Wachsmuth [2].

B. The Magnetic Field

The mass dispersion coefficient of an inhomogeneous field is given by

$$D = \frac{r_m}{1-n} \cdot \frac{\Delta m}{m}$$

where

$$n = -\frac{\partial B_z}{\partial r} \cdot \left(\frac{r}{B_z}\right) \qquad r = r_m$$

is the logarithmic field gradient and B_z the z-component of the magnetic strength. r_m is the radius of the medium trajectory. The coordinate system is shawn in Fig. 1.

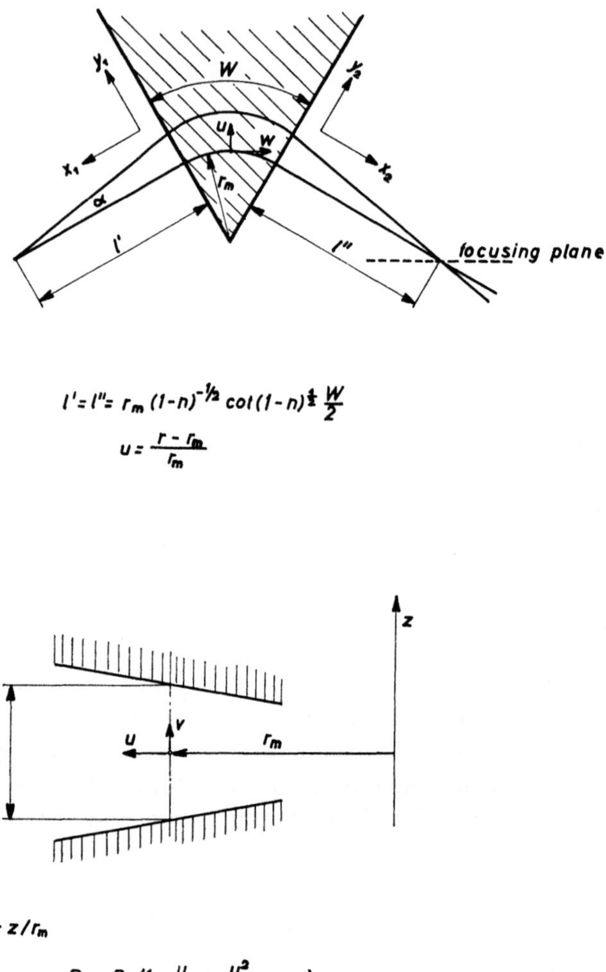

Fig. 1. Co-ordinate system of the magnetic Field

If we choose the medium radius $r_m = 1000$ mm and the inhomogeneity $n = \frac{1}{2}$, we get for an atomic mass 240 and for $\Delta m = 1$ a distance of 8.3 mm.

The object distance, which in our case is equal to the image distance, is given by

$$\frac{l'}{r_m} = \frac{l''}{r_m} = (1-n)^{-\frac{1}{2}} \cot(1-n)^{\frac{1}{2}} \frac{W}{2}$$

If we want to have focusing in the z-direction and second-order focusing in the radial direction too, we get the condition that the sector angle of the magnetic field must be $W = 169°42'$. The magnetic field then can be realized by conical pole faces [3].

Fig. 2 shows the axial component of the magnetic field strength in the medium plan. Because of the fringing flux at the inner and outer ends of the pole shoes the field strength deviates from the theoretical values. By a proper shape of the pole face this deviation can be corrected within the interesting region. The field measurements were done by means of two Hall-plates, one being fixed at the position of r_m, the other being moved. The object and image distances become $l = 816$ mm. Because of the fringing fields at the boundaries of the magnetic field, additional deflections of the ion beam will occur at the entrance and at the exit of the field. These deflections will be about 5° on each side. Thus the total deflection amounts to about 180°.

In Fig. 3, the ion trajectories are drawn for a radial opening angle of $\alpha_r = \pm 5°$ and a mass difference of $\Delta m = \pm 5\%$. In the z-plane the height of the object slit is 50 mm and the axial opening angle $\alpha_z = 1\frac{1}{2}°$.

Under the conditions mentioned above, the maximum width of the ion trajectories in the medium plane will occur at a sector angle of $w = 113°$. This width is 350 mm. The width of the ion beam is 143 mm at the entrance and 294 mm at the exit of the magnetic field. The beam has its maximum height of 112 mm at a sector angle of $w = 78.5°$. The beam height is 90 mm at the entrance and 16 mm at the exit of the magnetic field. For a momentum spread of $\beta = 10^{-4}$ and under the conditions mentioned above, the medium mass image broadenings due to the aberration of first and second order are as follows:

a) First-order chromatic aberration $\Delta y_\beta = \pm 0.4$ mm,
b) Second-order radial angular aberration $\Delta y_{\alpha_r^2} = 0$,
c) Mixed aberration $\Delta y_{\alpha_r \beta} = \pm 0.05$ mm,
d) Second-order chromatic aberration can be neglected, $\Delta y_{\beta^2} \approx 0$,
e) Second-order axial angular aberration $\Delta y_{\alpha_z^2} = 0$,
f) Agammatism $\Delta y_{\alpha_z \delta} = \pm 0.13$ mm,
g) Image curvature $\Delta y_{\delta^2} = -0.39$ mm.

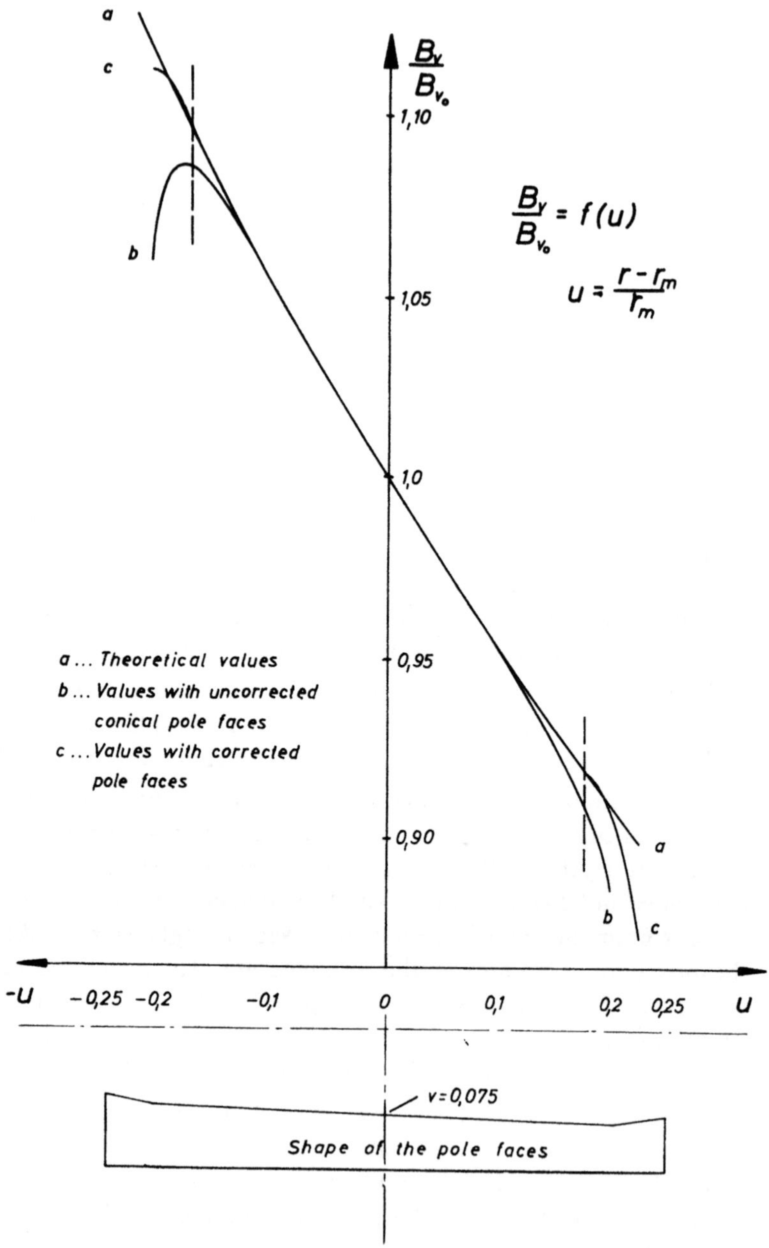

Fig. 2. Calculated and measured values of the axial magnetic field component

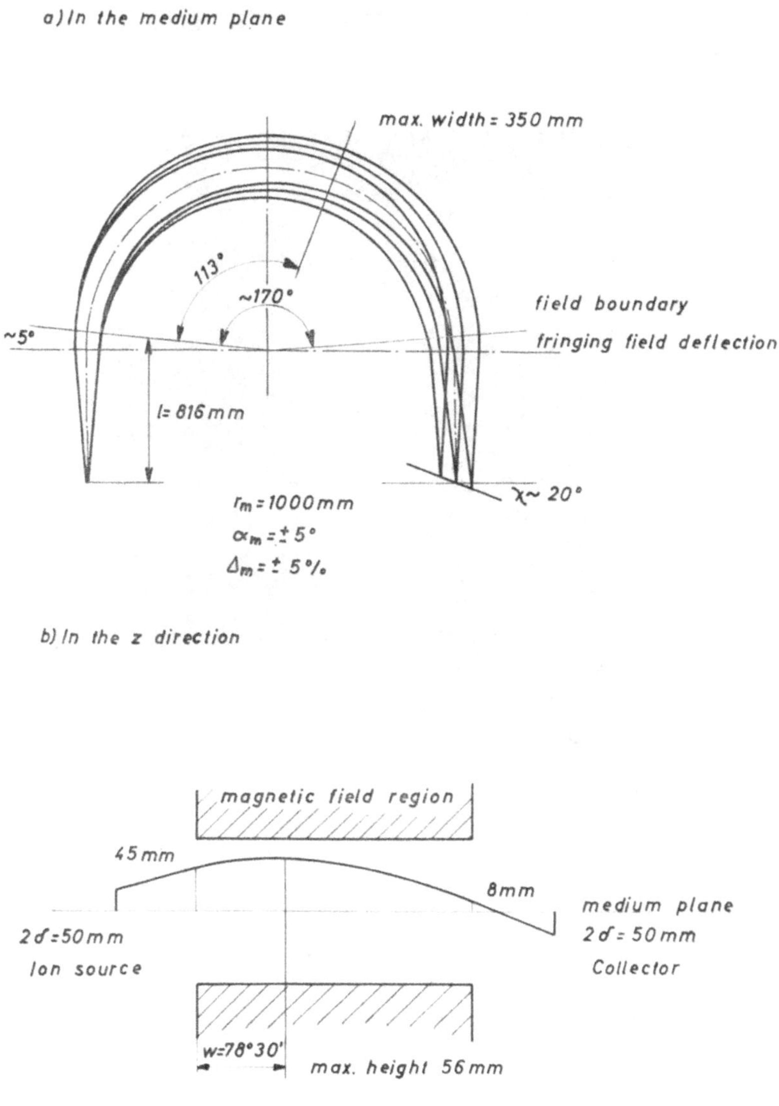

Fig. 3. Schematic ion trajectories

Therefore, the radius of curvature $R_{\mathrm{im}} = -650$ mm. The negative sign of R_{im} means that the centre of curvature has a negative y-co-ordinate. Due to the fringing fields we get an additional image curvature with the radius $R_{fr} = -500$ mm. This yields a total image curvature with a radius $R_{tot} = -280$ mm.

Fig. 4. Model of the magnet

Our magnet has the form of a "C" and the yoke is made of soft cast iron, the pole pieces of wrought iron. A view of the magnet model is given in Fig. 4. The diameter of the core is 1000 mm, the inner diameter of the pole pieces 1500 mm, the outer diameter 2500 mm, the total height 1600 mm. The distance between the pole faces at the radius of the medium trajectory is 150 mm. The maximum field strength at this place will be 5000 Oe. The coil will have 100,000 ampere turns, i. e. 2500 turns and a maximum current of 40 Amps. The power consumption of the coil will be nearly 20 kW. The copper turns are in a container and are cooled by oil. The maximum temperature of the oil when leaving the coil will be 40° C. The total weight of the magnet including the coil is about 20 t.

C. The Ion Source

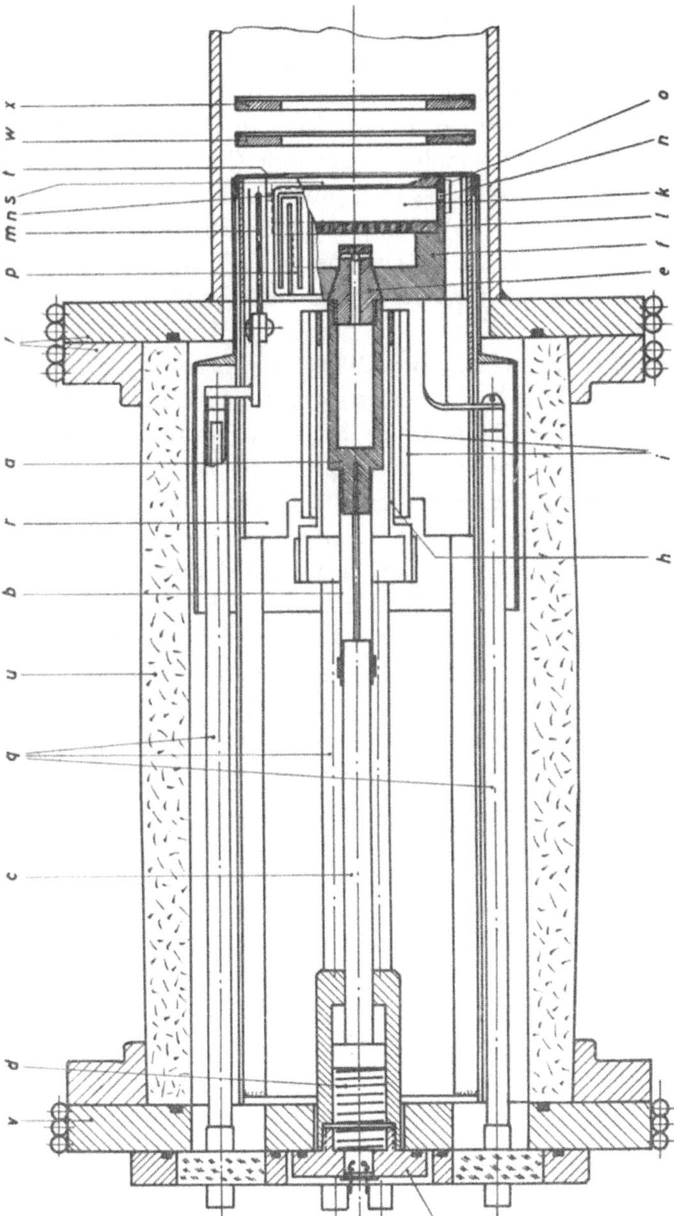

Fig. 5. The ion source

Details of the design of this source can be seen on Fig. 5. The first ion source we shall use is a conventional arc discharge source with an electron collimating magnetic field perpendicular to the ion beam. The charge material will be loaded in the graphite furnace (a). The furnace is made from pure reactor graphite and if desired, the chemically prepared charge material can be irradiated in the reactor together with the furnace. By a thin molybdenum tube (b), the furnace (a) is connected with a mobile stainless steel tube (c). By means of a spring (d) the conical front piece (e) of the vapour furnace (a) is pressed against the conical hole in the arc chamber box (f) which is also made of graphite. The vapour furnace can be loaded and unloaded with the aid of four screw bolts on the flange (g). The vapour furnace (a) is heated up to temperatures above 1000° C by means of a cylindrical graphite heater (h). This heater consists of a cylindrical graphite tube with a wall thickness of 2 mm which has six longitudinal slits along its circumference. The power consumption of the furnace will be 2 kW. Two cylindrical heat radiation shields (i) of molybdenum are arranged around the heater (h) to suppress the dissipation of heat. The vapour atoms entering the arc chamber (k) through a hole in the conical front piece (e) of the furnace are distributed over the entire chamber by means of a graphite distributor plate (l). The electrons emitted from a tungsten filament (m) and collimated by a magnetic field enter the arc chamber (k) through the hole (n). A repeller electrode (o) is provided to keep the electrons oscillating in the arc chamber. Additional heating of the arc chamber box (f) is possible by means of the two graphite heater plates (p). The conducting tubes (q) of the filaments as well as of the furnace and of the chamber heaters are water-cooled. The arc chamber and vapour furnace region of the ion source is surrounded by stainless steel shieldings (r). In front of the ion extraction slits (s) there is a molybdenum heat shield (t) in the form of a Pierce lens. The ion source unit is at a potential of +50 kV against ground and is insulated by a porcelain tube (u). The flanges (v) both at high potential and at ground potential are water-cooled. The ion extraction electrode (w) is at a potential up to −30 kV and is situated at a distance of 15 mm in front of the source slit. This negative electrode is water-cooled and can be manipulated from the outside of the vacuum chamber in order to facilitate the calibration of the ion beam. The distance between the negative electrode (w) and the grounded electrode (x) is 10 mm. The distance between the ion source shieldings (r) and the inner wall of the vacuum chamber will be quite small. This ion source is primarily designed to be used for solid charge material, but a gas inlet system will also be provided. It should be mentioned that the type described above is the first ion source we shall use. Later on experiments with other types of ion sources will be carried out, especially with a two-oven type for separation of small quantities of very short-lived isotopes.

D. The Vacuum System

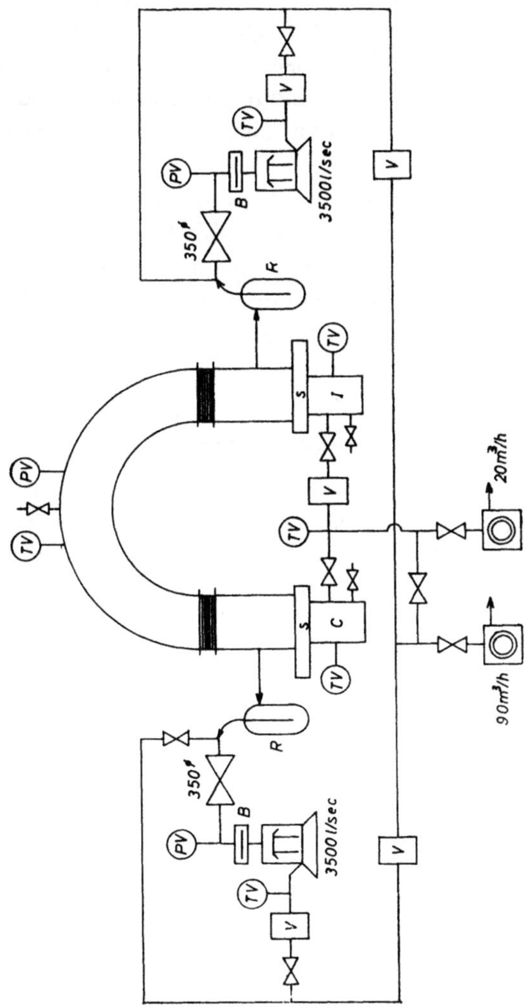

Fig. 6. The vacuum system

R Refrigerating trap; *PV* Penning vacuummeter; *TV* Thermoelectr. vacuumm.; *C* Collector; *I* Ion source; *S* Slide valve; *B* Water-cooled baffle; *V* Condenser vessel loaded with molecular filter

The ion source and also the collector are separated from the main vacuum system by electropneumatic slide valves and can be evacuated independently. Therefore, when loading or unloading the source or the collector material, we can keep the main vacuum chamber under high vacuum and special heating of the chamber should not be necessary. A schematic diagram of the vacuum system can be seen in Fig. 6. The pumping connections are situated in the space between the boundary of the magnetic field and the slide valves of the ion source and the collector, respectively. On the ion source side as well as on the collector side there is an oil diffusion pump with an unbaffled pumping speed of 3500 l/sec. Liquid air-cooled vapour traps are arranged between the oil diffusion pumps and the vacuum chamber. Most of the radioactive vapour particles will condense in these traps. The design provides easy changing of the refrigerating parts of these traps.

In the rough vacuum lines of the diffusion pumps there are containers with a molecular filter. If necessary this filter can be changed too. The valves of the vacuum system are also remotely operated from the control panel. The main vacuum chamber is made of welded stainless steel or aluminium. A liner of thin aluminium sheet prevents the contamination of the vacuum chamber walls.

E. The Power Supply

The power supply of the magnetic coil will consist of D. C. generator driven by a three phase synchronous motor. The current is stabilized $1:5 \cdot 10^{-5}$ by means of a balance amplifier. The signal for the amplifier input originates by comparing the voltage drop on a temperature stabilized series resistance with a Weston cell.

The positive as well as the negative high voltage is produced by a six phase selenium rectifier unit. The positive one will be stabilized by a series vacuum tube to $1:10^{-4}$. The power supply of the ion source, which is on high potential, comes via a 15 kVA insulating transformer.

F. The Plant Arrangement

A view of the electromagnetic separator plant is given in Fig. 7. To the left at the entrance to the separator room we have the control panel and the electronic racks. The floor of the separator room itself is 80 cm below the control room level to provide enough space for manipulating with a 1 t crane. The magnet on the right stands on a concrete base 50 cm high. Annexed to the separator room there is the machinery room. Here we have the positive and negative high voltage units, the power supply for the ion source, the heat exchanger

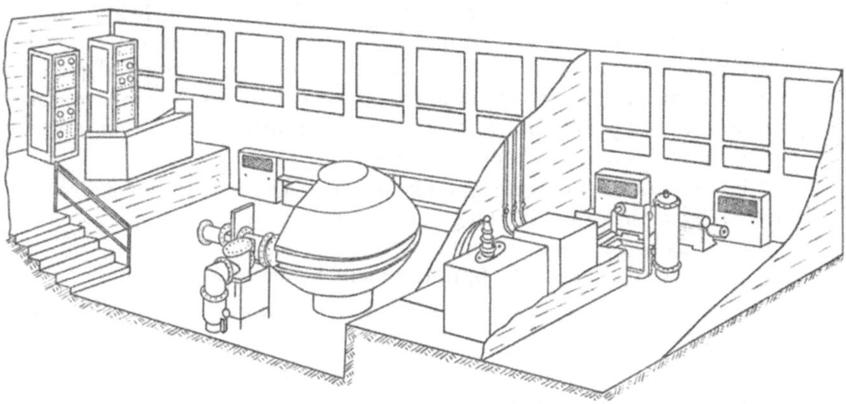

Fig. 7. The E.M. separation plant arrangement

for the magnet coil, and the backing pumps. In the channel on the collector side of the machine we have the cooling and rough vacuum lines going from the machinery room to the separator. In another channel on the ion source side of the machine we have the high voltage and power supply cables for the ion source. On the wall near the ion source we have reserved space for a lift to the chemical laboratory which is directly above the separator room. The maximum carrying capacity of the floor of the separator room is 2 t/m^2, therefore we have the possibility to build up lead shieldings around the separator.

The author would like to express his appreciation to all the co-workers of this project, especially to Dipl.-Ing. Buschbeck for the electronic work and Dr. Knizak for field measurements and model-experiments.

References

1. Svartholm N. and K. Siegbahn, Rev. Sci. Instr. **19**, 594 (1948).
2. Tasman H. A. and A. J. H. Boerboom, Z. f. Naturforschung **14 a,** 121 (1959).
   Wachsmuth H. and A. J. H. Boerboom and H. A. Tasman, Z. f. Naturforschung **14 a**, 818 (1959).

3. Svartholm N., Arkiv f. Fysik **2**, 115 (1949).
Boerboom A. J. H. and H. A. Tasman and H. Wachsmuth, Z. f. Naturforschung **14 a**, 816 (1959).

Discussion

R. W. McIlroy: Pile graphite may contain 0.1% of heavy metals, since its properties are determined by nuclear considerations only. For isotope collectors and any other use for which deposited material recovery is wanted a purer grade of material should be used. It is possible to obtain graphite with only 50 ppm of heavy metal impurity. The purer graphite may also have better machining properties.

F. P. Viehböck: The pile graphite we will use in the ion source and in the collector has the following specification:

Boron content 10 ppm
Ash content 50 ppm (Siemens Plania type AMT)

L. O. Love: Why are you choosing molybdenum for the heat shielding in the ion source?

F. P. Viehböck: Because of the great heat resistance of this material.

L. O. Love: A number of years ago we were using molybdenum sheets for shielding, but found that the material fused together and lost much of its insulation value. We substituted thin sheets of graphite, which seems to us to be more effective especially at temperatures above 1500° C. Also, if charge material is to be recovered from the ion source it is usually much easier to process graphite than metal. Which masses do you plan to process in your machine?

F. P. Viehböck: It is planned to process nearly all elements with a mass number above 20.

Plans for a Radioisotope Separator at CERN

By

G. Andersson

The Gustaf Werner Institute for Nuclear Chemistry, Uppsala

B. Hedin and **G. Rudstam**

SC Division, CERN, Geneva

With 5 Figures

The radioisotope separator in construction for the nuclear chemistry group at CERN is primarily intended for sample preparation in connection with spallation and high energy fission studies using the 600 MeV synchrocyclotron and the 25 GeV proton synchrotron.

This rather special application led to the following requirements on the performance of the separator:

1. Low intensity ($I_{tot} \lesssim 500$ μA) but high efficiency and reasonable dispersion up to the highest atomic masses; that is, the general characteristics of the Scandinavian type separators.

2. Simultaneous collection at equivalent focusing conditions of masses within ±10% from that of the central beam. This is necessary when both the stable carrier isotopes used as mass markers and far from stable radioisotopes are to be handled simultaneously. Even larger mass differences than the total span of isotopic masses of any element can of course be involved if the mass marker is not of the same element as the nuclides that are of interest.

3. In direct deposition approximately circular samples are desirable because end-window counters are used in the yield measurements, and scintillation spectroscopy will be used for the characterization of the nuclear species. A magnetic lens spectrometer may later be included in the equipment.

The third requirement calls for some sort of two-directional focusing. It is known that this can be achieved by adding an electrostatic cylinder lens system to the ordinary sector type separators or by using an inhomogeneous magnetic field. An essentially different and in principle very simple arrangement [1], however, turns out to have interesting properties in the present connection. It is based on the vertical focusing action[1] of the fringing field when an ion beam passes non-perpendicularly to the edge of a homogeneous field magnet. General formulae for this type of focusing have been given, for instance, by CROSS [2]. The special case to be considered here results if the incoming beam is parallel and enters the magnetic field at right angles (Fig. 1).

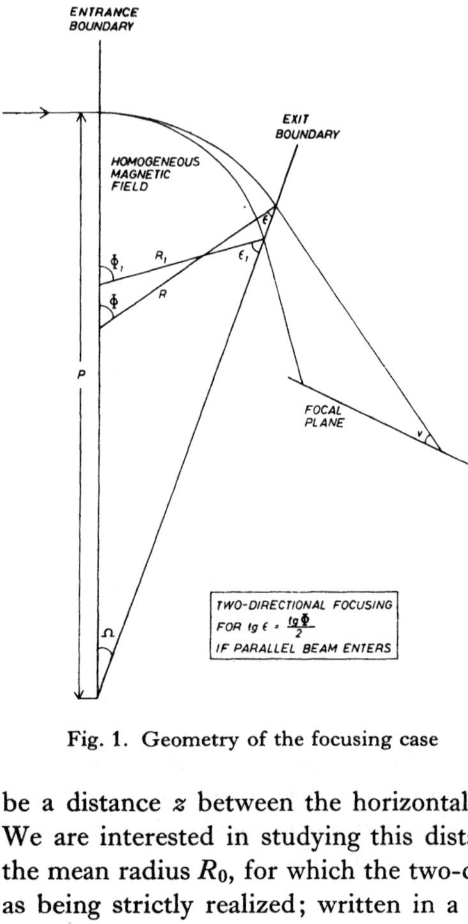

Fig. 1. Geometry of the focusing case

It is seen that coinciding horizontal and vertical focusing occurs for a very simple relation between the angle of deflection Φ and the angle ϵ between the exit boundary and the radius of curvature R in the exit point. With a straight boundary, however, this condition will not be generally fulfilled for other combinations, for instance Φ_1 and ϵ_1 in the figure. That is, there will be a distance z between the horizontal focus and the vertical focus. We are interested in studying this distance in the neighbourhood of the mean radius R_0, for which the two-directional focusing is assumed as being strictly realized; written in a series expansion

$$z(R) = z(R_0) + \left(\frac{\partial z}{\partial R}\right)_0 \Delta R + \frac{1}{2}\left(\frac{\partial^2 z}{\partial R^2}\right)_0 (\Delta R)^2 + \cdots$$

Using the proper formulas [2] and the notations of Fig. 1 we obtain

$$z = l_h - l_v = R\left(\frac{\cos\Phi\,\cos\epsilon}{\sin\Omega} - \cot\epsilon\right)$$

[1] Throughout the paper the main plane of motion of the ion beams is assumed to be horizontal.

The four parameters R, Φ, ϵ and Ω are connected by the relations

$$\sin\epsilon = \frac{P-R}{R}\sin\Omega \quad (P \text{ is kept constant})$$

$$\Phi = \epsilon + \Omega$$

Consequently, two of the parameters are independent and can be used to minimize $z(R)$ by making

$$z(R_0) = 0$$

$$\left(\frac{\partial z}{\partial R}\right)_0 = 0$$

The only significant solution of this system of equations turns out to be

$$\Omega = 19.48°$$

$$\epsilon = 35.27°$$

and thus $\Phi = 54.75°$.

This means, that the magnetic field must have the sector angle 19.5° and the central beam be deflected about 55° in order to achieve two-directional focusing with highest accuracy over a large mass difference. We shall refer to this focusing case as the 55° type.

The distance from the magnet boundary to the focal plane is given by $2R\cot\Phi$. As to the dispersion perpendicular to the beam, it is the same as for a symmetric sector analyser of the same radius. The focal plane, however, forms an acute angle with the beam direction. From geometrical considerations this angle is obtained as $v = 24.15°$ for the central beam.

In order to study the properties of this focusing case further, we assume that the magnetic field has been perfectly realized and let other sources of error, such as gas scattering and Coulomb repulsion, be represented by an opening angle α of the incoming, circular beam. A straight-forward application of general formulae [2] leads to the following expression for the horizontal and vertical line broadening, respectively, at right angles to the beam:

$$x = \frac{\alpha \cos\epsilon}{\sin\Omega}$$

$$y = \alpha \cot\epsilon$$

If we insert the extreme value of ϵ corresponding to the $\pm 10\%$ mass difference, or $\pm 5\%$ difference in radius of curvature, we find a variation in the line width of -4.5% and $+3\%$ relative to the central beam and a variation of the line height of -12% and $+11\%$. Further, the ratio x/y for the central beam is obtained $= 1.73$, implying that the focal spot will be an ellipse with the major axis horizontal. Within the $\pm 10\%$ mass difference the ratio of the axes is found to vary from about 1.6 to about 1.9.

It is generally recognized that the line broadening due to gas scattering etc. increases with increasing path-length of the ions through the separator. With ordinary sector magnets this is an argument in favour of a large angle of deflection. Thus a comparison of the present focusing system, which has an unusually small deflection angle, with the conventional types of apparatus is of some interest. It is seen from Table 1, that the 55° type in fact compares favourably even with the 180° machines in this respect.

Table 1. *Design parameters for* $D^{250} = 0.4$ *cm* (at right angles to the central beam)

Type of separator	Defl. angle Φ	Outfall angle ϵ	Mean radius R (cm)	Mean path length L (cm)	"Figure of merit" $D_A \cdot A/L$
Parallel beam	90	0	200	364*	0.27
Ditto	55	35.5	100	286*	0.35
Ditto	80	71	100	225*	0.44
Calutron	180	0	100	314	0.32
Sector	90	0	100	357	0.28
Ditto	60	0	100	450	0.22

* Distance ion source to magnet put = 50 cm.

The main problem in realizing the focusing principle is certainly to approximate sufficiently well the ideal magnetic field that has been assumed in the above discussions. In particular the exit boundary is of decisive importance. It may be remarked, first, that the small value of ϵ in this case is favourable. But even so, the actual lines of constant field strength as indicated in Fig. 2a tend to curve in such a way as to make the angle φ between the normal to the field strength line and the beam direction decrease with decreasing field strength. To fulfil the focusing conditions, however, φ has to be constant and equal to ϵ. Thus a reduction in focusing force and an increase in aberrations will result with a conventional magnet design.

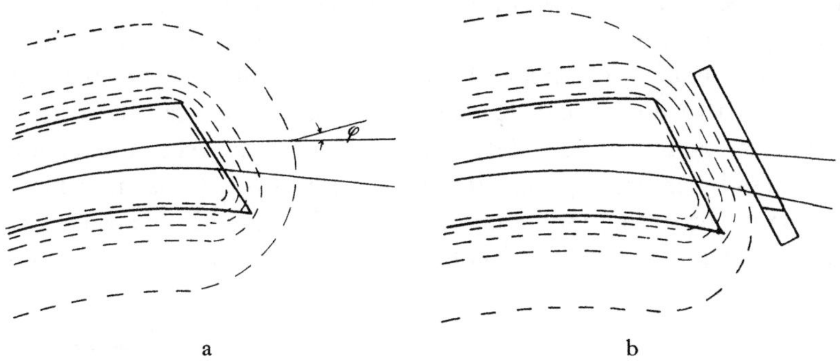

Fig. 2. Lines of constant field strength at a magnet edge
a. with conventional design b. with iron flange inserted

Fig. 2b demonstrates how the situation can be considerably improved by the insertion of an iron flange at some distance from the pole pieces. In Figs. 3a and 3b, flux lines and equipotential lines are drawn to show how the extension of the stray field is reduced when the iron flange is present.

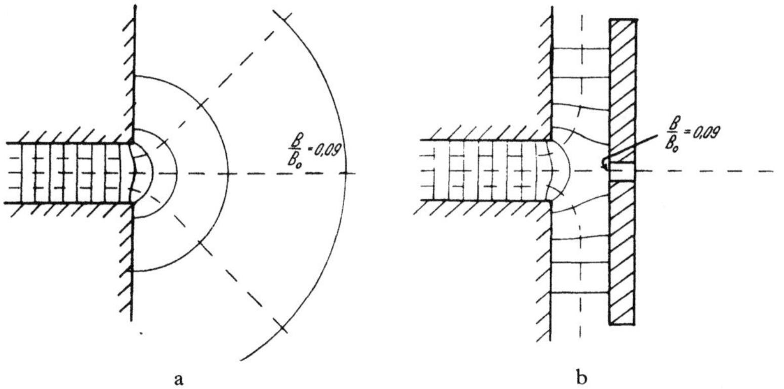

Fig. 3. Magnetic flux lines and equipotential lines at a magnet edge
a. with conventional design b. with iron flange inserted

Exact calculations of the proper position and dimensions of this shim are hardly feasible. Experience at CERN with similar cases of stray field correction were positive enough, however, to warrant the design of a separator based on this focusing principle.

The general lay-out of the apparatus is shown in Fig. 4. A spherical lens system, similar to the one in Uppsala [3], is used to form a parallel beam of the ions from a two-oven source of the Stockholm type [4], kept at max. 70 kV. The cylinder lens system also contained in the accelerating chamber of iron should really not be necessary with the two-directional focusing magnetic analyzer. It is to be considered mainly as a safety measure but may receive increased usefulness by an arrangement that makes it possible to use it either in the vertical or in the horizontal direction. The side of the accelerating chamber facing the magnet has been shaped for stray field reduction.

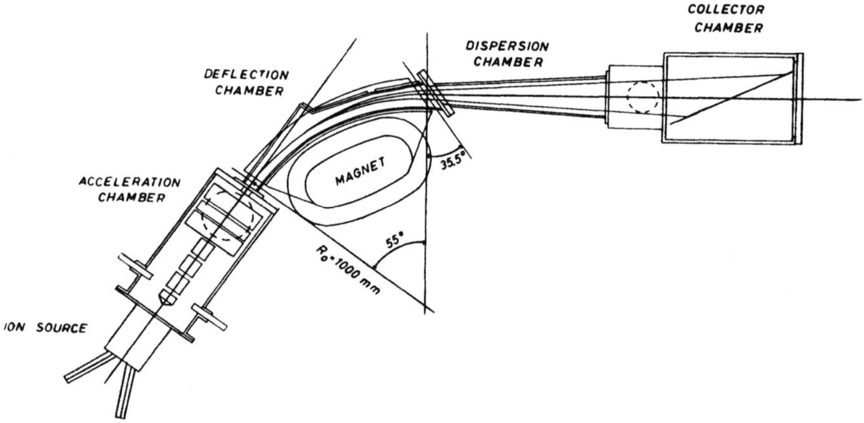

Fig. 4. General lay-out of CERN radioisotope separator

The mean radius of curvature was chosen as 1000 mm, corresponding to a dispersion of 4 mm perpendicular to the central beam at $A = 250$. As the beam will always be nearly parallel and have a diameter not exceeding 30 mm when it enters the magnetic field, the width and height of the gap can be kept low. The magnet has been designed with a gap height of 60 mm, while the width of the pole pieces increases gradually from 120 mm at the entrance to about 200 mm at the exit. The total weight of the magnet will be about 1 ton only and the power consumption for a maximum field strength of 5000 Oe will be well below 10 kW. The iron flange for field correction at the exit boundary forms part of what is here called the dispersion chamber. The collector chamber, finally, has been given considerable dimensions relative to the rest of the apparatus. This is partly a result of the narrow angle between beam direction and focal plane, but is also intended to allow activity detectors, including proper shielding, to be placed inside the chamber.

References

1. ANDERSSON, G., Ark. Fys. **16**, 483 (1960) (abstract).
2. CROSS, W. G., Rev. Sci. Instr. **22**, 717 (1951).
3. ANDERSSON, G., Ark. Fys. **12**, 331 (1957).
   BURGMAN, J. O., and G. ANDERSSON, Nucl. Instr. **3**, 33 (1958).
4. UHLER, J., and T. ALVÄGER, Ark. Fys. **14**, 473 (1958).

Discussion

CH. CASSIGNOL: Don't you fear that you will get purely geometrical focusing defects because of the oblique exit of the magnetic field?

G. ANDERSSON: As I said, the main problem is to realize the proper field form. We hope to achieve that by inserting iron flanges. But I don't know if we can get exactly the field shape we need.

CH. CASSIGNOL: I think it is a fact that for a sector machine of either 60° or 90° the geometrical defects are very small. I do not know why. They are in almost all cases smaller than the theoretical values. I do not know what they become in the inhomogeneous sector field or in the case of oblique entrance or exit of the beams with respect to the boundaries of the sector field.

B. HEDIN: I noticed the problem with bending magnets for high energy particles. At oblique entrance and exit of the beam the experimental focusing action in the vertical direction was only ⅓ of what was calculated. By shimming with iron flanges from both ends the experimental focusing quality improved to something like 90% of the calculated value. So it looks as if it is just a matter of shimming the edges properly.

Remark

By H. WAGNER and W. WALCHER on "Two Directional Focusing" (presented by W. WALCHER)

I should like to mention another way of two directional focusing. Dr. ANDERSSON reported that by choosing the appropriate angle between the boundary of the magnetic field and the central path two-directional focusing may be obtained by means of the fringing field. Care must be taken, however, to keep the fringing field undisturbed.

Two-directional focusing within the field may be obtained by using spiral ridge focusing which is also applied for high energy particle accelerators. Fig. 5 shows the pole piece into which spirally shaped grooves are milled. (If the width of the pole pieces is small as against the radius r, straight grooves may be used instead of spirally shaped ones.) The field strength in the middle plane of such a field is in first approximation given by the equation

$$B=\overline{B}\,[1+f\cos p(\vartheta+\delta)]$$

where $r\cdot\dfrac{\mathrm{d}\delta}{\mathrm{d}r}=\tan\gamma$ and p the number of the ridges on the circumference. In this field the equations of motion are

$$\frac{\mathrm{d}^2\xi}{\mathrm{d}\vartheta^2}=-\nu_r^2\,\xi;\qquad \frac{\mathrm{d}^2 z}{\mathrm{d}\vartheta^2}=\nu_z^2\cdot z$$

where $\xi=\dfrac{r-r_s}{\overline{r}}$; $r_s=r_s(\vartheta)$ is the radius of the equilibrium orbit and $\overline{r}$ a medium value of $r_s(\vartheta)$ along the circumference. The frequencies of the "Betatron oscillations" are defined by the equations

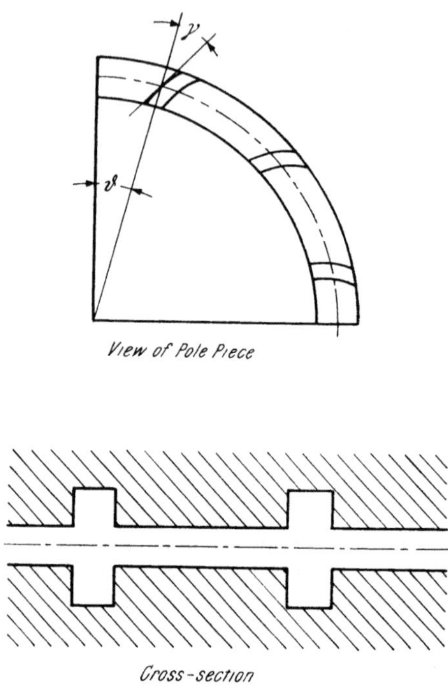

Fig. 5. Spiral-ridge field

$$\nu_r^2=1+\frac{\overline{r}}{\overline{B}}\frac{\partial\overline{B}}{\partial\overline{r}}$$

and

$$\nu_z^2=-\frac{\overline{r}}{\overline{B}}\frac{\partial\overline{B}}{\partial\overline{r}}+f^2\left(\frac{1}{2}+\mathrm{tang}^2\gamma\right)$$

$\dfrac{\overline{r}}{\overline{B}}\dfrac{\partial\overline{B}}{\partial\overline{r}}$ is the usual field index n. By choosing a constant field $B=\text{const.}$, n becomes 0, and hence $\nu_r^2=1$, $\nu_z^2=f^2\left(\dfrac{1}{2}+\tan^2\gamma\right)$. By choice of suitable values of f and γ one can realize the condition $\nu_z^2=1$, that means, one can obtain two-directional focusing. With $\gamma=72°$ which should be realizable, f becomes 0.3.

A field of this kind furnishes point imaging. The optical constants of the field (focal distance, situation of the cardinal points, etc.) are the same as those of the homogeneous sector field.

Ion-Optical Considerations on E.M. Separators

By

W. Walcher

Universität Marburg

With 1 Figure

We have considered how to construct the best isotope separator. Every isotope separator can be described as an optical system (Fig. 1). Particles of an energy ε (refraction index $n = \sqrt{\varepsilon}$) leave the meniscus of the plasma of the ion source (height y_Q) in the angular range $0 < u < \frac{\pi}{2}$. They are focused by an immersion lens (Pierce system or not) to an overcross.

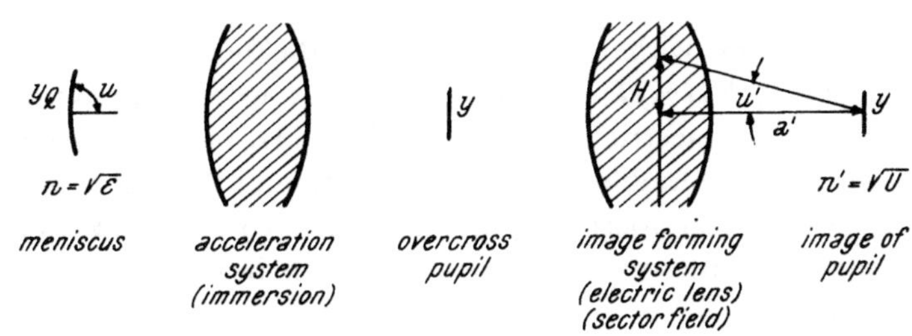

Fig. 1. The Electromagnetic Isotope-Separator as an Optical System

This overcross corresponds to the pupil in geometrical optics. The space charge effect, broadening the beam, may be described as a diverging lens contained in the immersion system. The pupil y is imaged into the collector plane by an electric and a magnetic lens. The image of the pupil has the height y', the aperture of the bundle in the image plane is u', the distance of the image from the back principal plane of the imaging system is a', the refractive index $n' = \sqrt{U}$.

For the formation without aberration of an image y' of the object y Abbé's sine condition must be satisfied. It can be shown that for the relation of corresponding values of meniscus and pupil as well the sine condition holds in a very good approximation so that

$$n y_Q \cdot \sin u = n' y' \cdot \sin u'$$

Since $\sin u = 1$ and $\sin u' \approx H/a'$

$$H = y_Q \cdot \sqrt{\frac{\varepsilon}{U}} \, \frac{a'}{y'}$$

In order to obtain high intensity, y_Q has to be made large; in order to obtain high resolving power and high purity of the separated isotopes y' has to be made small. Both demands lead to a great height of incidence, H, in the lens. A great value of H, however, means great spherical aberration ($\sim H^3$).

While this error is corrigible in magnetic fields a correction of the spherical aberration in electric lenses is hardly possible. (One can only choose from the known lens systems the one with the smallest spherical aberration.) The theoretically calculated electric lenses of smallest spherical aberration cannot be realized. We shall give a report on the "best" electric lenses for the purpose in question in the near future.

In a separator with a parallel ion bundle between electric and magnetic field, $H_{el} = H$; in an arrangement where the bundle diverges behind the electric lens, $H_{el} < H$. The first named arrangement, therefore, is unfavourable, as far as diminution of the resolving power due to the spherical aberration of the electric lens is concerned. The arrangement becomes the more favourable, the nearer the electric lens is to the overcross.

The Electric Quadrupole Mass Separator

By

F. v. Busch, W. Paul, H. P. Reinhard, and **U. v. Zahn**

Physikalisches Institut der Universität Bonn

With 4 Figures

A. Introduction

The quadrupole mass separator uses only a high frequency field for separating isotopes. Since a number of new experimental details appeared in the course of an investigation of this separator, we would like to present them, even though we have no special experience with radioactive isotopes.

B. Theory

The quadrupole field is produced by four hyperbolae at a distance $2r_0$ (see Fig. 1), to which a constant voltage U and an alternating voltage $V \cdot \cos \omega t$ are applied. Ions which are injected into the axis are caused to oscillate under the influence of the alternating field. These oscillations may be described by solutions of MATHIEU's differential equation

$$\ddot{x} + (a + 2q \cdot \cos 2\xi)x = 0$$
$$\ddot{y} - (a + 2q \cdot \cos 2\xi)y = 0$$

where $\omega t = 2\xi$

$$a = \frac{8eU}{m\,\omega^2 r_0^2}$$

$$q = \frac{4eV}{m\,\omega^2 r_0^2}$$

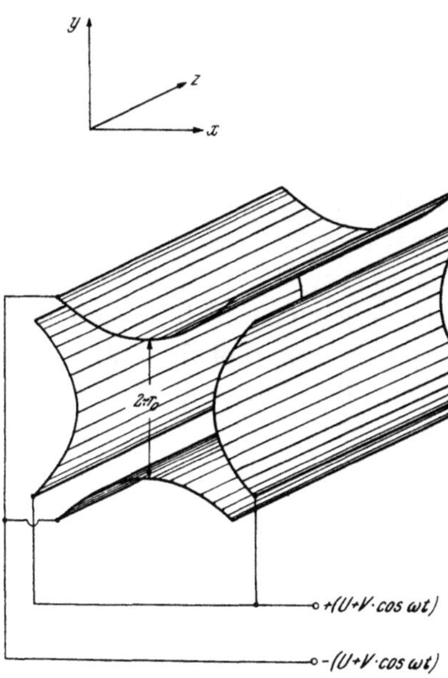

Fig. 1

(e = elementary charge, m = mass of the ion). The solutions are either "stable" or "unstable". In the latter case the amplitude of the oscillations grows exponentially with time, so that the ions strike the electrodes and are separated out. According to theory, the stability behaviour of the ions is given by the two parameters a and q. Only if the (a,q) value of an ion lies within a certain area (Fig. 2), does the ion pass

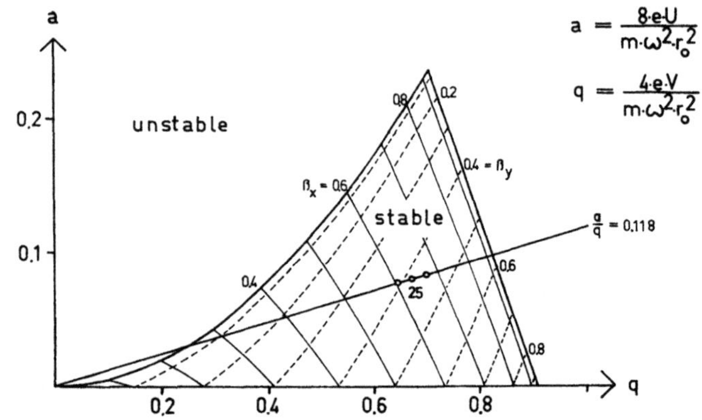

Fig. 2

through the mass filter. In this connection, it is important to point out that this (a,q) value does not depend on the initial conditions of the ions. Since the ratio a/q, which is equal $2U/V$, does not depend on the mass, ions of different mass all lie on a straight line passing through the origin of the stability diagram, if all field parameters are constant. By increasing the ratio U/V the stable mass interval may be made so small that only ions of a specified mass can pass through. All other ions strike the electrodes and do not pass through the filter. In this way one can attain a very high resolving power, up to 10,000, but only with very small currents. Unfortunately the separation of *large* ion currents at the peak of the stability plot is not possible, since near the (a,q) value at the stability limit the amplitude of the stable ions remains mathematically stable, but is so large that most of the stable ions, as well as the unstable ions, reach the electrodes. This results in a sharp decrease in transmission. Therefore, in order to separate large currents we place the operating point well within the stable region, for example $a=0.08$, $q=0.7$. Under these conditions the mass filter has a very high transmission but insufficient resolving power for isotope separation. This situation is corrected in the following manner.

The paths of the stable ions are described by the following equation

$$x = a_I \sum_{s=-\infty}^{\infty} c_{2s} \cdot \cos\left(s + \frac{\beta}{2}\right) \omega t + a_{II} \sum_{s=-\infty}^{\infty} c_{2s} \cdot \sin\left(s + \frac{\beta}{2}\right) \omega t$$

As is evident, the fundamental frequency of the ion motion is $\underset{s=0}{\omega} = \frac{\beta}{2} \cdot \omega$, where β is a parameter that is different for the x and y components of the motion but otherwise is determined solely by the (a,q) value (see Fig. 2). As a result different isotopes of an element have different fundamental frequencies within the stable region. If one now applies an auxiliary alternating field in the form of a homogeneous field whose frequency is the same as the frequency of an isotope, then this isotope is in resonance, and its amplitude grows linearly with time. Small auxiliary field strengths are sufficient to separate out an isotope entirely. Neighbouring isotopes are excited to beats with an amplitude which decreases with decreasing field strength and increasing mass difference. This principle has been extended in our separator by the application of two auxiliary frequencies to separate out two or more isotopes.

C. Apparatus

A magnetically stabilized low voltage arc serves as the ion source. The energy inhomogeneity of the ions does not affect the operation of the mass filter. The ions are drawn out of the ion source by an extraction lens and are injected directly into the quadrupole field. This is produced by four brass rods, which gives sufficient approximation to hyperbolae. The collector is a smooth piece of aluminium foil perpendicular to the field axis. The vacuum in the field region need not be very good since the strong focusing prevents small angle scattering from having an influence on the resolving power.

At the beginning of April 1960 we performed our first large scale separations in which we attempted strong enrichment of the rarer isotopes magnesium 25 and 26. The dimensions of the apparatus and some experimental parameters are:

Ion source (magn. stab. low voltage arc):

arc current	0.4... 1 A		
arc voltage	20... 60 V		
magnetic field	300...600 gauss		
exit hole diameter	5 mm	voltage	+1.5 kV
extraction lens diameter	6 mm		−1.6 kV
entrance diaphragm diameter	20 mm		+1.5 kV

Quadrupole field:

length	3 m
radius r_0	1.5 cm
frequency $\omega/2\pi$	2.6 Mcps
ac amplitude, V	2.6 kV
dc voltage, U	175 V
hf power	400 W

Auxiliary fields:

frequency $\omega_1/2\pi$	883 kcps (eliminating Mg^{24})
frequency $\omega_2/2\pi$	798 kcps (eliminating Mg^{26})
amplitude of both	~80 V

Collector:

aluminium foil (uncooled)

total current of $Mg^{24,25,26}$ without auxiliary fields
0.8 mA

current of Mg^{25} (rel. abund. 10%) with auxiliary fields
80 μA' (uncorrected for secondary electrons)

yield of Mg^{25}	30 μg/h
purity	see Table 1

Pressure: $1\ldots5 \cdot 10^{-5}$ mm Hg
(with 2 oil diffusion pumps of 100 l/s)

Total energy consumption:

ion source + electronics	2.0 kW
3 pumps	1.0 kW
	3.0 kW

Table 1. *Isotope Separations*

Sample			Feed	I	II	III
Auxil. fields	ν_1	kcps		789	798	832
	ν_2	kcps		922	883	883
	V_1	Vp		25	35	42
	V_2	Vp		68	55	55
Duration	t	h		10.5	11	9
Current without sep.		μA		600	720	840
Current during sep.		μA		92	50	83
Weight of samples		μg		394	>140	265
Analysis	Mg^{24}	%	79.0	3	9	3
	Mg^{25}	%	10.0	32	50	13
	Mg^{26}	%	11.0	65	41	84
Mass wanted				25	25	26

We have found it difficult to make a continuous check of the separation factor. It depends primarily on V, U, ω, ω_1 and ω_2. If one requires the separation factor not to change over many hours, then one has either to construct very highly stabilized electronics or to accept the lower long-time stability and provide a method of checking the separation factor in the course of a run. For this purpose we have provided a small hole in the collector foil through which a small part of the collector current can pass, for analysis, into a second small quadrupole field of 5 mm field radius and 50 cm length. In this way it is possible to measure the relative abundances of the isotopes on the collector and to adjust the separator for maximum separation. In this connection one problem arises which has not been solved satisfactorily at the present time. If the resolving power is too low or the field parameters are incorrect, then contamination by the neighbouring isotopes on the collector depends on the position on the collector. Thus, having a fixed hole in the collector, we only get an approximate value for the instantaneous purity.

Space charge leads to a shift in the a-value, which is different for the x and y components. It is

$$\Delta a_y = -\Delta a_x = \frac{127 \cdot I_{[\mathrm{mA}]}}{r_{0[\mathrm{cm}]}^2 \cdot \nu_{[\mathrm{Mcps}]}^2 \sqrt{A \cdot U_{[\mathrm{V}]}^{\mathrm{acc}}}}$$

and with

$$|\Delta a_{\max}| \approx 0.1 \qquad r_0 = 1.5\ \mathrm{cm} \qquad \nu = 2.6\ \mathrm{Mcps}$$

we get with the low accelerating voltage of $U^{acc} = 1.5$ kV a maximum current of 2 mA, i.e. above this value the transmission and resolving power of the mass filter decreases because of space charge. Compensation of this effect by electrons is not possible since these are extremely unstable by virtue of their small mass. It seems possible to obtain a certain amount of compensation by changing the field radius, but this has not yet been tried.

If one wants to increase the efficiency of the mass filter further one must first increase the current and simultaneously also the resolving power of the auxiliary fields. Both increases could be achieved if it were possible to construct relatively long quadrupole fields. For this purpose rods are unsuitable for use as electrodes and therefore we have built a quadrupole field in which the electrodes are formed by groups of stretched wires. The wires are fastened only at the two ends of the field. The field is 8.2 m long, vertical in position and precise enough so that we have achieved a resolving power at the top of the stability diagram of 10,000. Up to now we have not intended this apparatus for the separation of large currents. On the other hand, this method of field construction can be directly applied to isotope separators. This technique has the advantage that there is not one parameter in the entire apparatus which has to be adjusted mechanically when the field is replaced. The position of the ion source, as well as the position of the collector, with respect to the field axis, is uncritical since the entrance diaphragm is *not* focused on the collector like an optical image. Furthermore, because of its lower capacity per meter, the wire system requires approximately 35% less hf power than the conventional rod construction.

References

1. Paul, W., and H. P. Reinhard, Forsch.-Ber. d. Wirtsch.- u. Verkehrsmin. Nordrhein-Westf., Nr. 450.
2. Paul, W., and H. P. Reinhard, Symposium über Isotopentrennung (Amsterdam 1957).
3. Paul, W., H. P. Reinhard and U. v. Zahn, Z. f. Phys., Bd. 152, 143—182 (1958).

Discussion

M. J. Higatsberger: From the solution of Mathieu's equation one can see that only (a/q) values are important for the stability of a given mass. I do not quite see where the restriction in the current as mentioned should come in if only U and V are the mass determining factors.

U. v. ZAHN: In order to reach the detector an ion must have a maximum amplitude lower than r_0, the field radius. In spite of a stable working point this maximum amplitude depends on the initial conditions (angle of entrance, distance from axis and velocity) with which the ion begins to oscillate in the quadrupole field. Therefore for higher resolving power or purity of samples one needs to reduce the radius of the ion beam entering the quadrupole field.

M. J. HIGATSBERGER: If the amplitudes become too great, can this be corrected by using different r_0's along the path?

U. v. ZAHN: Theoretically this is possible. But in this case the voltages must be altered along the path too, for the ratio U/r_0^2, which is proportional to a, and V/r_0^2, which is proportional to q, should be kept constant along the field.

R. W. MCILROY: Have any cost comparisons been made between this method and ordinary electromagnetic separation?

L. O. LOVE: We made a study and came to the conclusion that we should have considered the quadrupole separator very carefully if we did not already have calutrons. However, one might save power if a number of quadrupoles could be operated in parallel within a single instrument.

U. v. ZAHN: In the quadrupole field you have a high field strength in the region between the quadrupole electrode rods and the vacuum tank and in this region you also have a high power consumption (see Fig. 3). Now the system may be altered to use 9 rods with 4 mass filters (see Fig. 4).

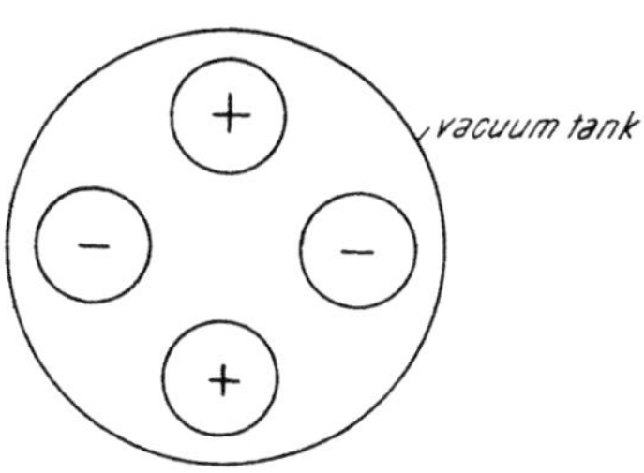

Fig. 3

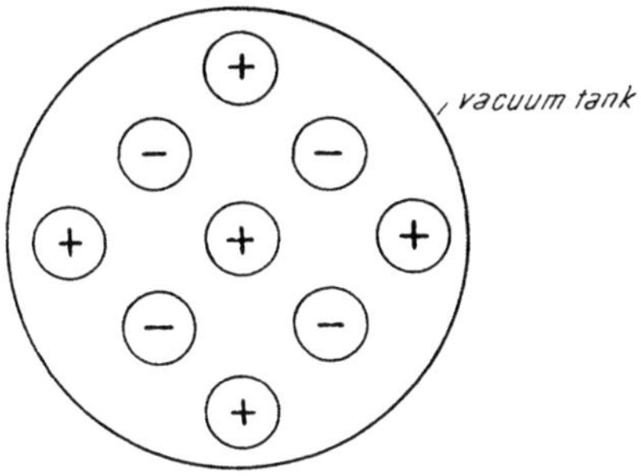

Fig. 4

In this case the power consumption is not 4 times but only 2.6 times the power consumption of 1 mass filter.

R. W. MCILROY: What considerations govern the length of the rods?

U. v. ZAHN: The length of the rods is important for the accelerating voltage. The desired resolving power gives the number n of high frequency periods, in which the ions have to stay within the field. n depends on the accelerating voltage, the frequency and the length L of the field. For a given accelerating voltage one must have a long field or have a higher frequency. The power consumption is $\sim C \cdot \nu^5 \cdot r^4 \cdot m^2$ where C is the capacity of the quadrupole field and ν the frequency. Therefore, it is more economical to have a longer field than to have a higher ($L \sim C$) frequency.

R. W. McIlroy: What limits the possible ion current?

U. v. Zahn: The upper limit for the current is given by the space charge effect and the length of the field. For a given length, frequency, and mass, there is an upper limit for the accelerating voltage which limits the current to be drawn out of the ion source.

H. J. Langmann: Could you not increase the effective length of your system by adding a static axial magnetic field?

U. v. Zahn: No. It has been calculated but adds no advantage.

Chapter III

Operating Experiences on E. M. Separators

Double Magnetic Deflexion Isotope Separator for the Production of Very High Purity Isotopes

By

R. Bernas, J. L. Sarrouy, and **J. Camplan**

Université de Paris, Laboratoire de Physique Nucléaire, Orsay (S. et O.)

With 11 Figures

Summary

The electromagnetic separator of the Nuclear Physics Laboratory of Orsay is made up of a 60° sector homogeneous magnetic field analyser followed by a semi-circular inhomogeneous field analyser of the Svartholm Siegbahn type. The enrichment factors reached are from ten to a hundred times higher than those obtained with other separator types while the ion beam intensity remains in the neighbourhood of 1 milliampere.

A description of the instrument is given as well as the first results obtained in the separation of isotopes of a few elements: Cr, Sr, Yb, Hg, U, etc.

I. Introduction

The production of separated isotopes in ponderable amounts, which developed extensively after 1945, has facilitated or made possible numerous studies in various fields from nuclear physics and optics to the most classical chemistry, thanks to the isotope dilution method. Stocks of enriched isotopes of nearly all the elements have been built up in various production centres (Oak Ridge, Harwell, Moscow) and are easily available.

However, the method of preparation of these isotopes in gramme amounts by electromagnetic separators has to a certain extent sacrificed purity in order to achieve quantity.

In numerous studies, it has now become necessary to make use of isotopes much purer than those available at present.

This is the case when two neighbouring isotopes have very different cross sections for the same nuclear reaction or when the ratio of their abundances is of the order of, or superior to, 100.

In nuclear spectroscopy, one must be able to separate clearly isotopes present in quite different proportions which emit the same type of radiation with very closely neighbouring energies. The situation is similar in optics for studies on hyperfine structures.

Finally, the stable isotope dilution method represents a very important field of use, and its sensitivity depends in large measure on the purity of the available isotopic tracers.

The preparation of very pure isotopes by recycling of a mixture already enriched is made practically impossible by the low efficiency of electromagnetic separators (1 to 5% except in a few semi-industrial cases, and then at high cost).

It has seemed more efficient to us, in order to achieve very high isotopic purity, to carry out double dispersion of the same ionic beam by two magnetic analysers. The principle of double magnetic deflection has already been applied in mass-spectrometry by Inghram and Hess in 1953 [1] and by White and Collins 1954 [2], in order to determine with high accuracy the abundances of certain rare isotopes. In electromagnetic separation Cassignol [3] has achieved the first instrument incorporating two successive deflections: one electrostatic stage suceeding a classical magnetic one. He has obtained some interesting results in this way; however, in this machine, the sharpness of the lines is critically limited by the effects of the space charge in the second stage. Other reasons which we shall presently study, in addition to this one, have led us to prefer the double magnetic deflection.

The main causes of contaminations can be limited to the four following types:

1. Interactions between accelerated ions and residual gas molecules.
2. Defects in focusing resulting either from the ion optics, or from an incomplete space charge neutralisation.
3. Inhomogeneity of ion energies.
4. Condensation in the collectors of neutral vapours issued from the source.

The studies made up to the time when we began our work did not permit us to specify with any accuracy which of the above processes was predominant. Only partial results had been given on very high output instruments [4].

Hence, we have essentially taken account of the fact that the general results and particularly the shape of the mass spectrum obtained with the first separator, built by one of us [5], in considering the effects of ion inhomogeneity as a secondary factor and our choice has been a system with two successive momentum filters.

The systematic study since made on the Saclay separator has confirmed our hypothesis at least in first approximation [6].

However, one can determine that the ultimate performances of such a machine would be limited by the following factors:

1. The presence of slowed down ions or of ions which have not been accelerated by the full voltage.
2. Rapid fall of the high voltage due to breakdown.

To this, one must add that two identical magnetic stages will not permit elimination of impurities coming from the aperture defects of the first stage. However, it is possible to eliminate the effects of the last two points, on the one hand by introducing an electrostatic condenser which will prevent ion beams from entering the second stage in case of breakdown and on the other hand, in choosing, as the second stage, an analyser with a higher dispersion than the first.

The choice of the characteristics of the second analyser, was dictated by the desire to realize simultaneously a high dispersion and a minimum length of ion trajectories.

These two conditions can be satisfied by the use of an inhomogeneous magnetic field of cylindrical symmetry, such as proposed by SVARTHOLM and SIEGBAHN [7].

Indeed, in this field, which is defined by the following relation:

$$H_{(r)} = H_{(r_0)} \left(\frac{r_0}{r}\right)^{\alpha} \qquad (1)$$

where $H_{(r)}$ is the intensity of the field on radius r
r_0 the mean radius
α the field index

The dispersion is given by:

$$D = \frac{1 + G_r}{1 - \alpha}$$

G_r, the radial magnification, is equal to unity in all cases where object and image are symmetrically disposed with respect to the analyser. Hence in this case, and when for example $\alpha = 1/2$, the dispersion is twice that of the homogeneous ($\alpha = 0$) field with the same radius.

Moreover, the field defined by relation (1) possesses axial focusing properties which it is desirable to use in order to reduce the width of the beam (along the Z axis), and hence increase the transmission of the machine.

The evident superiority of inhomogeneous magnetic sectors over the usual homogeneous sectors, may lead us to consider that they can be used for the first as well as for the second stage of an isotope separator. We must, however, note that the increased dispersion results in a decrease of the range of masses which one can simultaneously focus on a collector. Indeed the useful area of the pole pieces of an inhomogeneous field magnet is generally smaller than in a homogeneous one and, except in very special cases, it is desirable to collect a large range of masses in good condition at the first as well as at the second stage.

The last point to define now concerns the relative position of the centres of curvature of the two magnetic stages with respect to the ion trajectories. If, as a first approximation, the choice can be immaterial, the intermediate focus having been considered as the object of the second analyser, it seems preferable to employ arrangement A of Fig. 1 which effectively realizes the addition of the dispersion of the two stages and in this way permits the simultaneous collection of several isotopes at the final collector.

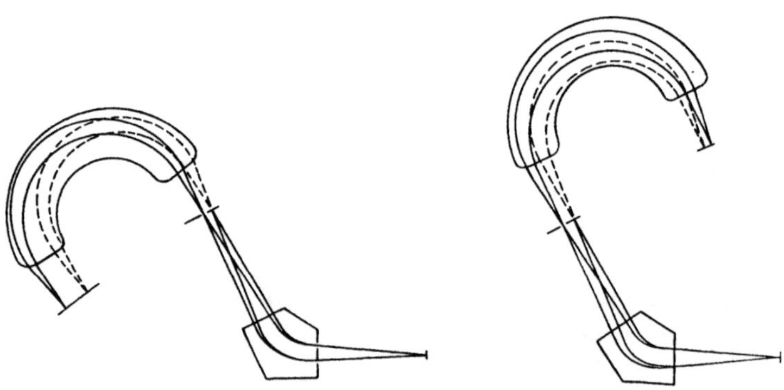

a b

Fig. 1. Two possibilities of connecting successive magnetic cascades

(a) The centres of curvature are situated on opposite sides of the trajectories. The resulting dispersion is the sum of the dispersions of each stage

(b) The centres of curvature are situated on the same side and the resulting dispersion is the difference between those of each stage

Moreover, in the separation of very heavy elements, when the width of the slit of the intermediate diaphragm may not be neglected with respect to the spacing of neighbouring isotopes, arrangement A would make it possible to achieve higher isotopic purities.

II. Description of the Machine

Fig. 2 shows a schematic diagram of the separator.

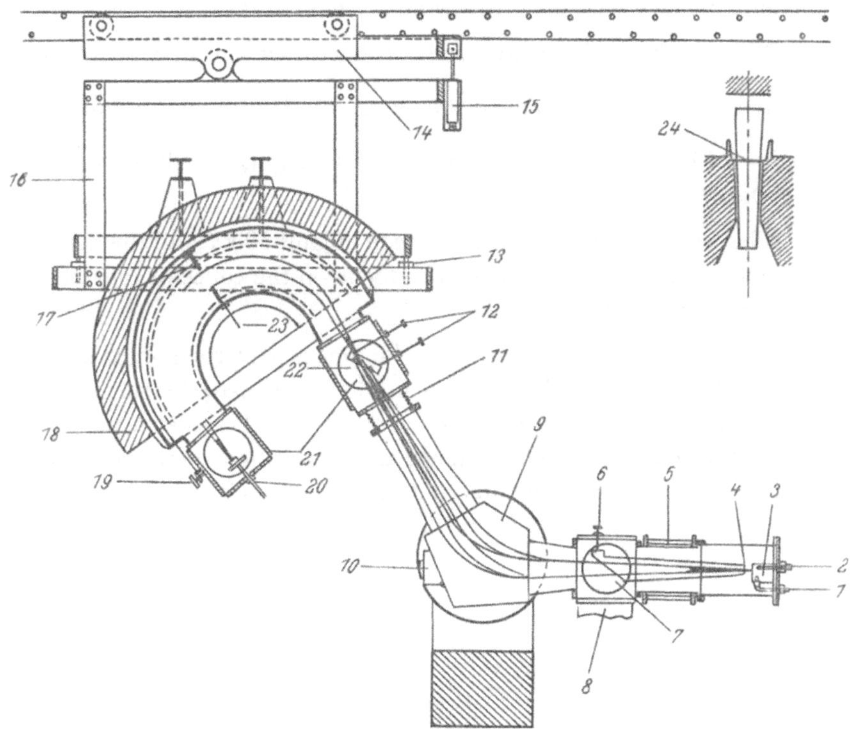

Fig. 2. Schematic drawing of the separator

1 Oven supply 2 Supply to filament 3 Source 4 Accelerating electrode 5 Insulators 6 Mechanism for electrode motion 7 Glass-window 8 Vacuum pump 9 Electromagnet pole pieces 10 Window 11 Metallic bellows 12 Device for positioning intermediate diaphragm 13 Vertical jacks 14 Carriage 15 Angular jacks 16 Electromagnet frame 17 Upper half diaphragm 18 Electromagnet 19 Device for motion of upper half-diaphragm 20 Final collector 21 Window 22 Pump 23 Lower half-diaphragm 24 Metallic spacers for increasing strength of vacuum chamber

The first stage is a 60° homogeneous magnetic sector field with a curvature radius of 50 cm. It was built by the firm Gamma Industry in Paris following the drawings of the machine built by one of us (R. B.) in 1953 at the Commissariat à l'Energie Atomique and which is at present working in Saclay.

The fact that this first stage is identical to that of Saclay, to which an electrostatic stage [3] has been added since, allows comparison of the two types of second stages and makes also possible interesting complementary studies of the same phenomenon.

The ion beam intensity at the collector of the first stage varies from 1 to 2 mA according to the elements. The resolving power $\frac{M}{\Delta M}$ defined with respect to the width at half height, is 400 for these intensities.

The acceleration voltage is between 30 and 50 kV and the magnetic induction in the gap is sufficient to focus on the collector ions of mass 250 accelerated with 50 keV.

The extraction and focusing of the ions is performed by means of a single electrode carried by a very simple stainless-steel structure. This electrode may be adjusted under vacuum, which makes it possible to adjust the source electrode distance as well to orientate the accelerating electrode along the horizontal and vertical axes (Fig. 3).

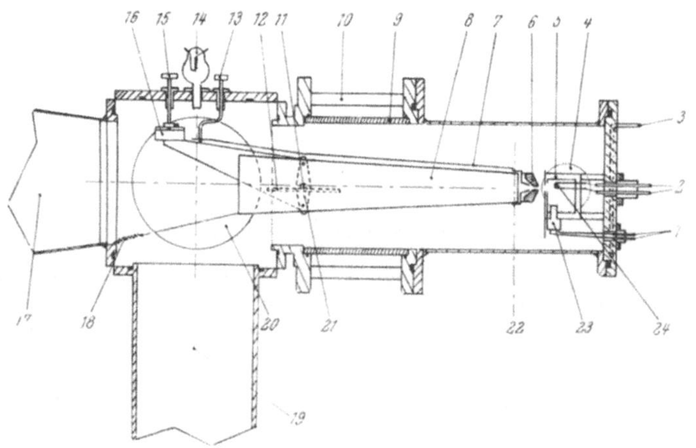

Fig. 3. Region of the ion source and of the accelerating electrode

1 Supply to oven 2 Supply to filament 3 Cooling water 4 Source magnet pole pieces 5 Source 6, 8 Accelerating electrode 7, 13 Gear for rotation of accelerating electrode 9 Pyrex insulation 10 Insulating columns 11, 12, 15 Device for axial motion of electrode 14 Ion gauge 16 Counter weight 17, 19 Parts of the vacuum chamber 18 Tension spring 20 Window 21 Electrode frame 22 Axis of rotation of the tip of the accelerating electrode 23 Oven and crucible

The region of the collector of the first stage is made up of a light-metal cube (Fig. 4) with six 20 cm diameter holes. It is possible in this way to install observation-windows, mechanical devices or

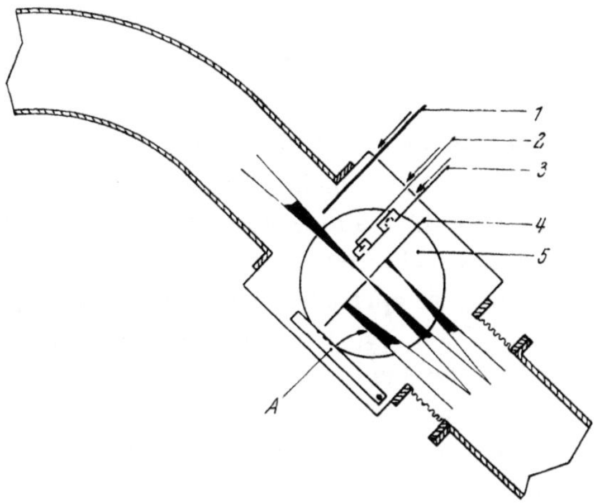

Fig. 4. Schematic drawing of the collector and intermediate diaphragm

A Vacuum tight lock for collector 1 Stop plate to shut the entrance to the second analyser 2, 3 Movable collectors for the mass spectrum measurements or the deposition on filament or crucible 4 Intermediate diaphragm 5 Window

liquid air traps if necessary. A 1500 l/sec oil diffusion pump is connected at this point. The intermediate diaphragm is adjustable in position along 3 perpendicular directions and its surface may be covered with thin metal foils, easily interchangeable, when separating radioactive isotopes. The coupling of the two magnetic stages was determined by the fact that the pole faces of the first magnet were in a vertical plane, hence we suspended the second analyser on two steel beams under the ceiling of the hall. Stresses between the two stages are eliminated by connecting the two vacuum chambers with stainless steel bellows of large diameter.

The second stage: the preliminary theoretical studies of the second analyser which we mentioned above, had led us to consider the construction of an inhomogeneous field sector of radial symmetry, with the index $\alpha = 0.5$.

For reasons of simplicity, we ordered from the Swedish firm K.B.C.M. a magnet with characteristics very close to those we desired, which had been built for studies on nuclear spectroscopy for C. Mileikowsky [8].

According to a project of Svartholm [9] this magnet has conical pole faces and gives a field of the shape:

$$H_{z(r)} = H_{z(r_0)} \frac{2r_0}{r + r_0} \qquad (2)$$

This relation is equivalent to (1) of the first order when $\alpha = 1/2$.

The sector angle is 180°, which, when one takes into account the fringing field, corresponds to a deflection of 191.4°. The magnetic gap is 52 mm for $r = r_0$ and is limited by two circles of 330 and 470 mm radius. The radial angular aperture is close to 3°; it is defined by an adjustable aperture diaphragm situated half way between the intermediate focus and the final collector. The position of this focus which acts as the object for the second analyser, is at a distance of the order of 30 cm from the entrance face of the magnetic field and the image obtained is 22 cm away from the other face. The intermediate diaphragm and the collector can then be handled easily and observed through large windows.

The radial aperture aberration, in the case of inhomogeneous sectors, has been computed by several authors [10]. For the arrangement which we have used, it has a theoretical value close to 0.2 mm which is quite acceptable, considering the dispersion of the instrument.

The position of the intermediate focus being liable to vary slightly as a function of the different parameters (intensity, interelectrode distance etc.), mechanical devices allow us to place the intermediate diaphragm and the collector in the positions corresponding to the best focus.

The vacuum chamber

It is drawn schematically on Fig. 2. The sides of the chamber are made up of two stainless steel sheets 3 mm thick welded to 2 circular bands of radius 280 and 600 mm with heights of 43 and 57 mm.

This design has the advantage of making the best use of the magnetic gap and of allowing a high pumping speed in the region of the beam despite the small space between the pole pieces. The rigidity of the chamber is achieved by the use of stainless steel spacers situated outside the region through which the beam passes (24, Fig. 2).

The two half diaphragms situated in the centre of the chamber can be adjusted from the outside when the machine is working. The rapid variation of the index due to the fringing field would lead to important aberrations unless the extent of the beam in the radial direction is limited.

The chamber is fixed on the magnet yoke by adjustable supports which will facilitate the mounting and demounting of the chamber. Pumping is done at both ends of this chamber, the 1500 l/sec pump near the intermediate diaphragm being assisted by a second oil diffusion pump near the final collector; each pump is followed by a liquid air trap. Access to atmospheric pressure of the collector of each stage is made possible through the shutter A (Fig. 4) and B (Fig. 5).

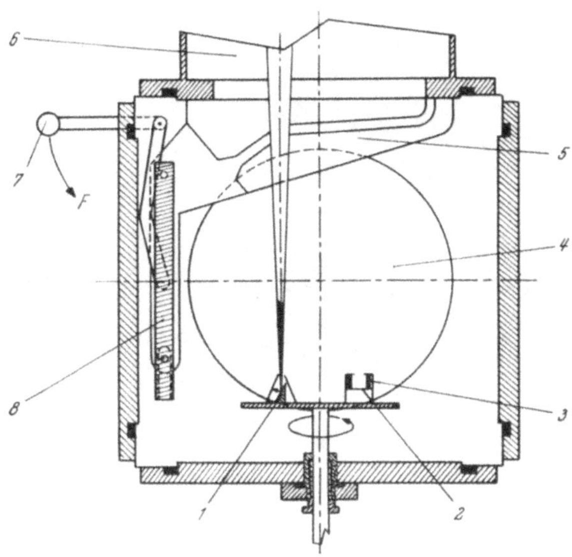

Fig. 5. Final collector

1 Collector pocket for high collection efficiency 2, 3 Insulated plates for the measurement of the mass spectra (180° rotation will allow mass spectrum recording of the beam in the same position) 4 Window 5, 7, B Quick action vacuum lock for closing of collector chamber

Supplies and electronic stabilizers

The high tension set delivers 100 mA at 50,000 V and is stable to 1/5000 [5]. It is continuously adjustable.

The two electro-magnets are supplied separately. The first has a resistance of 30 ohms and is excited by a 15 ampere current. For the second, these values are 0.1 ohm and 600 amperes respectively. Both magnets are supplied by motor generator sets. Electronic circuits for the stabilisation are classical[1] and the results of stability tests made during 30 minute periods indicate that the variations are smaller than $\frac{1}{10{,}000}$; drifts are very small and can easily be compensated by the operators. The high tension supply, that of the ion source and the primary pumps, is situated in a room directly below the separator.

Fig. 6. Ion source for low vapor pressure elements (operating temp. 1700°C) To the left in the source the filament, to the right the electron reflector, to extreme right the source slit and the front electrode

The control panel is shown on Fig. 7.

[1] That of the first stage has been described earlier (5), that of the second stage is the work of M. Cabrespine of this laboratory.

Supplies to the various parts of the ion source as well as to the space charge neutralisation electrode are analogous to those described previously [5].

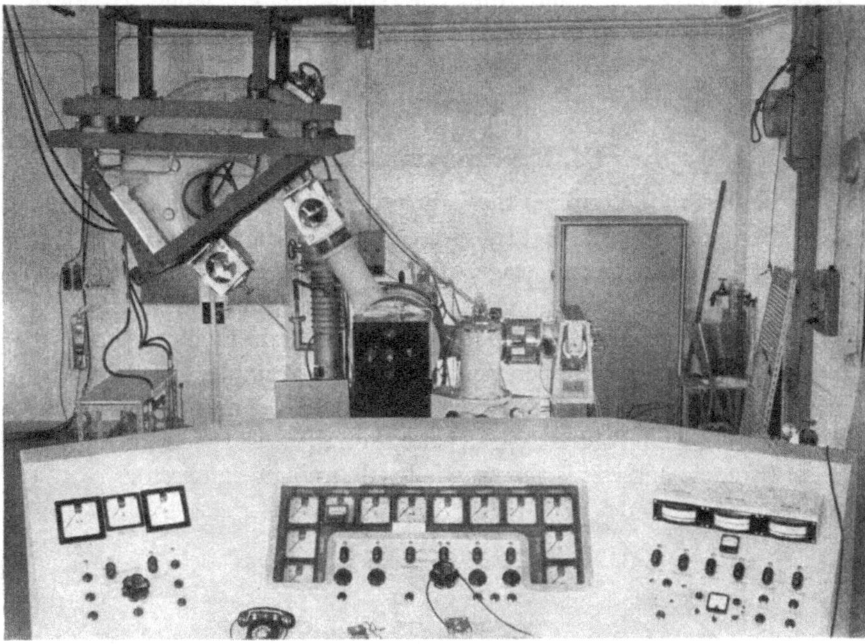

Fig. 7. Photograph of the isotope separator

Right to left: The ion source region set at 50,000 V; the pumping section, and the homogeneous field analyser of the first stage. The collector with the intermediate diaphragm, the inhomogeneous field analyser fixed on its carriage and the final collector with its adjustments. The control panel with its three parts: on the left, the electronic stabilisation units, in the centers control for the ion source and to the right vacuum control and measurements

Ions sources

The ion sources used are derived from those previously published [11]; they operate under normal conditions up to temperatures reaching 1700° C (Fig. 6) and are intended for the rapid introduction of radioactive samples [14]. The maximum output given by these sources reaches 10 mA, which permits the separation of stable isotopes in weighable amounts (1 to 2 mA at the collector at the first stage) and mass-marking can be done easily for identification purposes when separating radioactive isotopes. The high operating temperature of the ion source will reduce very much the time necessary to collect radioactive isotopes.

The efficiency of the separator defined as the ratio of the activity of one isotope introduced in the source to that measured after separation varies from 1 to 2%. An ion source of the duo-plasmatron type was tested recently [12] and the corresponding efficiency is more than ten times that of the usual sources; the xenon spectrum obtained under these conditions showed a very satisfactory resolving power. Tests on this source are being continued.

III. Experimental Results

In order to determine the advantages of the double-magnetic deflection we collected the isotopes of a few elements successively at the first and the second stage of the machine and this under the same conditions of operation (intensity, pressure, ion optics etc.).

The very high purity of the isotopes collected at the final stage makes it a very delicate chemical operation to prepare the samples for analysis by mass spectrometry. For this reason we used the technique of direct deposition of the ions of the desired isotope on the mass spectrometer filament (isotopic analysis is then made by thermionic emission).

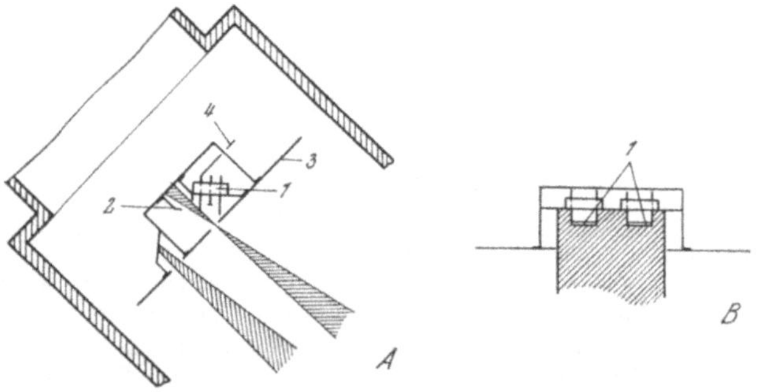

Fig. 8

A. Device for collection of an ion beam on a mass spectrometer filament
1 Filament assembly corresponding to the mass spectrometer ion source. The current to this filament is measured through lead 4 2 The ions not collected on 1 will deposit in this pocket
B. Relative position of the triple filament assemblies and of the collector pocket

Fig. 8 represents one of the arrangements employed which makes it possible to analyse either the deposit made on the mass spectrometer filament or the deposit made on a gold foil behind this filament. The spectrometer filaments used were either of tungsten, tantalum, or platinum, according to the elements, and were very carefully purified before introduction into the separator.

The quantities necessary to perform an isotopic analysis are obtained in a few minutes or in one to two hours, according to the natural abundance of the isotope and the nature of the element.

The results obtained for several elements are shown in Table 3.

Table 3. *Experimental results*

	Isotope coll.	Time of Collection	type	Isotopic Composition (%)						Enrichment factor
				50	52	53	54			
Chromium				4.3	83.76	9.55	2.38			
	Cr^{53}	10 h	C	—	0.1	**99.9**				8,800
		½ h	F	—	0.016	**99.984**	—			52,000
		½ h	F	—	0.005	**99.995**	—			>100,000
				84	86	87	88			
Strontium				0.56	9.86	7.02	82.56			
	Sr^{87}	10 min	F	—	—	**99.94**	0.06			~ 20,000
	Sr^{84}	48 h	C	96.1	0.56					4,000
					151	153				
Europium					47.77	52.23				
	Eu^{153}	6 h	C		0.02	**99.98**				~5,000
				170	171	172	173	174	176	
Ytterbium				3.03	14.31	21.82	16.13	31.84	12.73	
	Yb^{170}	1 h	F	**98.85**	0.363	0.257	0.150	0.307	0.069	1,300*
	Yb^{173}	1 h 30	F				**99.85**	0.117		1,700*
	Yb^{174}	13 h	C					**99.95**	0.035	~10,000–12,000
				198	199	200	201			
Mercury				10.02	16.84	23.13	29.80			Contamination
	Hg^{199}		C	0.2	**≥99**	0.5	0.2			by neutral vapour
					235	238				
Uranium					0.7	99.3				
	U^{235}	1 h	F		**99.0**	1.0				14,000
		1 h	F		**98.75**	1.25				11,000
	U^{238}	½ h	F		$\leq 5.10^{-6}$	**99.9995**				> 1,400
										Preliminary test

* Contamination by the natural element.
(°) C: deposit on a collector F: deposit on a filament

The lack of a sufficient supply for the second magnetic stage limited the tests during one year to those elements of mass below 100 and the results given in Tables 3 and 4 must be considered as minimum performances.

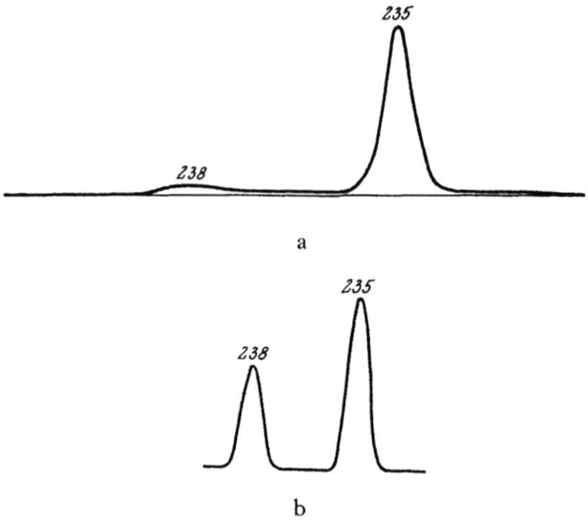

Fig. 9

(a) Mass spectrum of U^+ of the final collector

(b) Mass spectrum of the deposit at the collector of the first stage during the same run

The analysis after the 2nd magnetic stage of the ions which have gone through the intermediate diaphragm during the collection of U^{235} is shown in Fig. 9. The mass spectrum of the deposit obtained at the collector of the first stage during the same collection is indicated in that figure as a means of comparison.

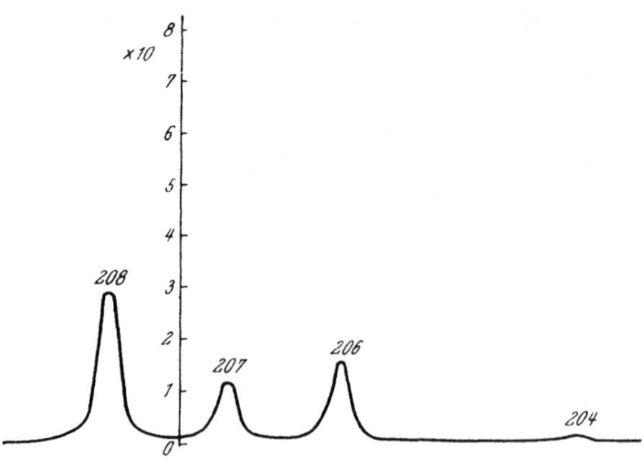

Fig. 10. Mass spectrum of lead at the final collector after the elimination of the intermediate diaphragm (in the case of simultaneous collection of several isotopes at the second stage)

Fig. 10 represents a mass spectrum of lead at the collector of the second stage when the intermediate diaphragm is suppressed. In fact.

in order to collect simultaneously several isotopes at the second stage it is preferable to cut into the intermediate diaphragm the slots corresponding to the desired isotopes so as to eliminate the background visible on Fig. 10 arising from ions scattered in the first stage.

The improvement brought about by the second magnetic stage is shown clearly in Table 4 which gives the values of enrichment factors obtained at the first stage and at the second stage under the same conditions.

The example of Yb^{173} shows that preferential contamination by heavier isotopes also occurs in a double magnetic deflection isotope separator.

Table 4. *Comparison of the enrichment factors obtained with one or two stages under the same experimental conditions*

Isotope studied	Type of Collection		Enrichment factor*		$\frac{\gamma_2}{\gamma_1}$
	1st Stage	2nd Stage	1st Stage (γ_1)	2nd Stage (γ_2)	
Cr^{53}	F	F	540 [1]	52,000–100,000	100–200
Sr^{87}	F	F	305	20,000	70
Yb^{170}	F	F	100	1,300	13**
Yb^{173}	F	F	92 [1]	8,000 [1]	87
Yb^{173}	F	F	47	1,700	38
U^{235}	F and C	F	200	14,000	70

* Defined with respect to the neighbouring isotope of heavier mass except [1] where it is mass M—1.

** Contamination by natural Yb responsible for this small value.

Separation of radioactive isotopes

The various elements of the separator lend themselves particularly well to the separation of radioactive isotopes. In the first place, the contamination of the second vacuum chamber can be reduced to a minimum by the presence of the intermediate diaphragm (which can be cooled to liquid nitrogen temperature) and by efficient pumping at this level.

The simplicity of the ion source, which does not have any high tension insulation, and that of the accelerating electrode structure will permit easy decontamination. The introduction of radioactive samples is carried out in one to two minutes through a vacuum lock in the ion source and the extraction of isotopic targets is possible in one minute [14]. In this way, we have repeatedly been able to separate mercury isotopes from a gold foil irradiated in the cyclotron;

the total time elapsing between the arrival of the foils near the separator and the moment when counting could begin was 3 to 4 minutes [13].

Table 5. *Radioactive isotopes separated at 1st and 2nd stage*

Hg^{197}	Hg^{195}	Hg^{194}	**197–188**: Simultaneous collection at 1st stage relative cross-section measurements
Hg^{192} Hg^{189}	Hg^{191} Hg^{188}	Hg^{190}	**192–188**: Individual separations at 2nd stage for β and γ spectroscopic studies
Hg^{187}	Hg^{186}	Hg^{185}	Direct spectroscopic studies on a scintillator at the collector of 2nd stage half-lives about 4 min, 80 sec and 50 sec
Au^{194}	Au^{195}	Au^{196}	Studies at 2nd stage (γ spectroscopy)

The separator allows the simultaneous collection of isotopes comprised in a range of $\pm 5\,\%$ about an average mass at the first stage and $\pm 2.5\,\%$ at the second stage. This technique has permitted measurement of the relative cross section of some spallation reactions [15].

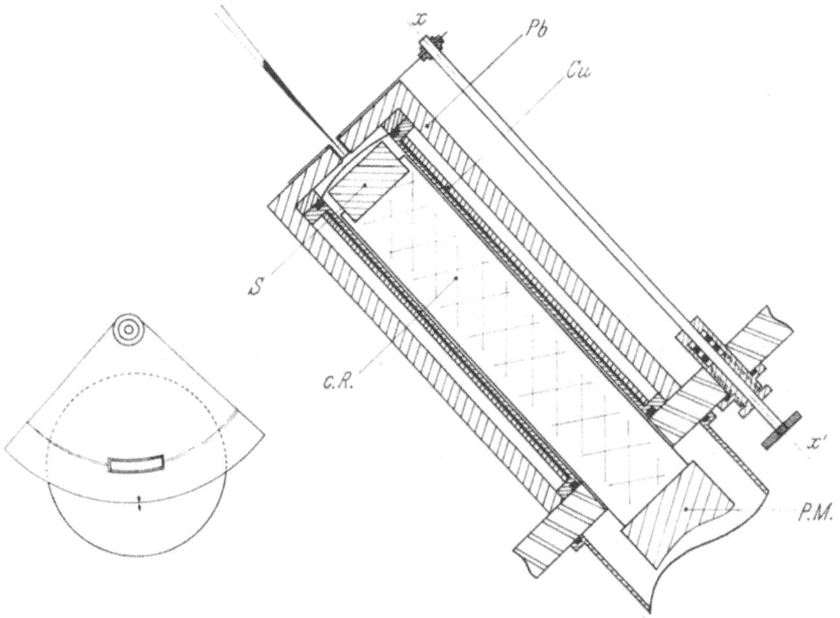

Fig. 11. Device for the direct study of the short-lived isotopes at the collector of the isotope separator

We have found it useful to replace the usual collector (20, Fig. 2) by a photomultiplier and light guide (Fig. 11) which makes it possible to study directly the radiation emitted by isotopes of short half-lives and weak activity, even in the presence of very active neighbours. This last characteristic is due to the possibility of stopping the unwanted isotopes on the diaphragm of the first stage (1 m away) and by introducing a lead screen between the two collectors. This device has been used in particular to detect new isotopes of mercury with short half-lives: Hg^{185}, Hg^{186}, Hg^{187} [16].

Discussion of the results – residual contamination

The results obtained indicate that double magnetic deflection presents very important advantages with respect to the types used previously. The enrichment factors are more than ten times higher than those of single magnetic deflection machines or of the mixed magnetic electrostatic cascade.

It may be interesting to compare to our results those obtained by this last machine for the same element and with approximately equal current at the collector (0.3 mA total at the first stage), and for the same duration.

The enrichment factors obtained for U^{235} for these two machines are, respectively, 500 (mixed cascade) [18] and 14,000 (magnetic cascade). The improvement brought about by the second magnetic deflection is here evident.

In practice, the isotopic purity which we have achieved will make it possible to solve most of the problems which arise at present, and in particular to furnish extremely pure tracers of a great number of elements for studies by the isotope dilution method.

However, important improvements may be obtained. Indeed, the study of Table 4 shows that the enrichment factors corresponding to this second stage alone are lower than those of the first stage, while the characteristics of the second analyser should result in higher values if the ionic contamination were due only to particles having the same energy.

For this reason we carried out a study of the residual contamination in the second stage. Detailed results will be published later [17] but the first conclusions it is possible to draw show that a not inconsiderable fraction of the contamination is due to ions which have an energy lower than that corresponding to the full acceleration voltage.

Although this conclusion is contrary to that of some authors [6] one must notice that it may have easily been overlooked until now, since it affects (in the case of uranium for example) a fraction of the order of 1/10,000 of the ions of the neighbouring isotopes.

The elimination of this amount would be sufficient to lead to an enrichment factor of 60,000, while the present one is only 14,000. This new value would correspond to a ratio γ^2/γ^1 of 300 which may be considered as a satisfactory performance for the second magnetic stage.

It is then apparent that it will be necessary to take into account new processes of contamination which have been neglected up to the present before predicting the limiting performances of the double magnetic stage machines. In particular it is not possible merely to multiply the enrichment factors of each of the stages operating separately as has already been proposed [18].

References

1. Inghram, M. G., and D. C. Hess, Nat. Acad. of Sciences, Nat. Research Council Washington D.C.: public. no. 311, 24 (1954).
2. White, F. A., and F. L. Collins, Appl. Spectroscopy 8, no. 4, 169 (1954).
3. Cassignol, C., Proceedings of the Symposium on isotope separation, Amsterdam 1957, p. 631 (North Holl. Publ. Co.).
4. T.I.D. 5217, Electromagnetic separation of isotopes. Technical Information Service U.S.A.,E.C.), no. 5217. Edited by Gutherie and R. K. Wakerling (1949).
5. Bernas, R., Phys. Radium **14**, 34 (1954).
6. Cassignol, C., Rapport C.E.A., no. 1185, C.E.N. Saclay (1959).
7. Svartholm, N., and K. Siegbahn, Ark. Fys. vol. 33 A, no. 21 (1946).
8. Mileikowsky, C., Ark. Fys. vol. 4, no. 16 (1952); vol. 7, no. 3 (1953).
9. Svartholm, N., Ark. Fys. vol. 2, no. 14, 115 (1950).
10. See H. Ikegami, R.S.I. **29**, 943 (1958).
11. Bernas, R., Réalisation d'un séparateur électromagnétique d'isotopes, thèse (Paris 1954).
12. Klapisch, R., Thèse 3ème cycle, Etude et réalisation d'une source à champ magnétique inhomogène; application à la séparation électromagnétique des isotopes (Paris 1959).
13. Albouy, G., R. Bernas, M. Gusakow, N. Poffe, and J. Teillac, C.R. Ac. Sci. **249**, 407–409 (1959).
14. Sarrouy, J. L., See page 184, present volume.
15. Poffe, N., G. Albouy, R. Bernas, M. Gusakow, M. Riou, and J. Teillac, Réactions (*p*, *n*) induites dans l'or par des protons de 155 MeV J. Phys. Radium **21**, 343 (1960).

16. ALBOUY G., GUSAKOW M., POFFÉ N., J. Phys. Radium **21**, 751 (1960).
17. BERNAS, R., J. CAMPLAN, and M. VAN MENTS, J. Phys. Radium (to be published).
18. Cf. ref. [6] p. 35.

Discussion

F. BUSCHBECK: You touched on the question of the stability of your supply. What did you have in mind with these figures: stability of the magnetic field, or of the acceleration voltage or of their ratio?

R. BERNAS: The instability concerned the second magnetic analyser's supply.

H. J. LANGMANN: Do you stabilize your supplies absolutely or do you use some sort of moving pin like the Swedish group?

R. BERNAS: The high voltage is stabilized better than one part in 5000 but there is no pin system at all.

P. K. ROL: Instead of the diaphragm in the 90° position of the second stage, is it not possible to use a curved entrance boundary of the magnet perhaps combined with a "HERZOG-Blende"?

R. BERNAS: Practically we find that we lose extremely little intensity on the diaphragm, we have a little insulated collector on it and we can measure the current to this diaphragm. When you deal with very high purity isotopes, even if a very small amount of impurity gets to the collector it would be very annoying. So the diaphragm was there to stop even small side beams or scattered ions but it does not stop much of the main beam itself.

F. P. VIEHBÖCK: Have you observed any great image distortion due to the fringing field at the 2nd magnetic deflection stage?

R. BERNAS: We have observed image distortion in many cases but we have not made studies to prove where they come from.

R. BERNAS: A question to Dr. J. H. FREEMAN: I wonder whether you observed a defocusing action in "Hermes" due to the fringing field effect at the entrance and exits of the magnet. As you know the fringing field will lead to a y-component and this component tends to spread part of the beam in the z-direction.

J. H. FREEMAN: We observed no defocusing in the z-direction.

W. WALCHER: Dr. HEDIN is interested in z-focusing. Maybe he could give some comment on it.

B. HEDIN: On the separator we are constructing at CERN, we try to avoid all these errors in the z-focusing by using iron plates close to the edge of the magnets to shorten the distance of the fringing field. This distance will be about 80% of the gap width. We do not know the results on a separator but we have good results with bending magnets for high energy particles.

R. BERNAS: We are also planning to put a magnetic short circuit at the entrance of the magnetic field. But one has to be careful. In our machine and also in Dr. Cassignol's the aberration due to angular divergence is well corrected without any stray field shields, so this situation may be altered if we use such shielding.

O. Skilbreid: Was the sharp corner of your coma defect at the light-mass side or at the heavy-mass side?

R. Bernas: We have observed it on the heavy-mass side.

Ch. Cassignol: We observed in the first stage of the conventional Nier-type instrument that the coma changed its sign when the intensity of the beam was raised.

O. Skilbreid: In the Copenhagen machine we have two deflecting plates before the entrance of the magnetic field. We have found that the coma is very much influenced by the potential difference between these plates. In fact, we very often obtain the best focusing conditions when the potential of these plates is about 1 kV, which would give a deflection of about 1°.

Electromagnetic Separation of Radioisotopes of the Heavy Elements at the Oak Ridge National Laboratory

By

L. O. Love, G. M. Banic, W. A. Bell, and **W. K. Prater**

Oak Ridge National Laboratory, Oak Ridge, Tennessee, USA

With 11 Figures

Introduction

Since the early 1940's some radioisotope separation has been intermittently performed in the calutrons located at Oak Ridge, uranium being the most notable. With minor modifications being made to the equipment, Th^{230} also has been separated. In 1952 one calutron system was modified to accommodate $PuCl_3$ charge and to provide also for the total containment of plutonium during the separation process. Thus, over the past few years, the isotopes of thorium, uranium, and plutonium have been separated on a recurring basis. Isotopes having much less activity, namely Be^{10}, Bi^{210}, K^{40}, Sm^{147}, etc., have been separated more on a "when needed" basis.

The original calutron and its performance in the separation of uranium and of some stable isotopes is generally well known. The performance of several modified versions of both ion source and receiver is likewise known. This paper discusses an ion source which was designed for use in plutonium separation but which has been sufficiently versatile that its usage has been extended to include also uranium and stable isotope separation. Also discussed is a receiver concept which appears desirable in "single isotope" collections; included are typical results attained in the various separations programs.

Ion Source Description

Plutonium containment imposes inherent difficulties in servicing such bulky components as are used in the calutron. In fact no servicing of plutonium-contaminated equipment was even contemplated in the inception of the plutonium program. During the period 1953–1957 all ion sources were used once, then washed and prepared for burial.

In 1957, the plutonium and heavy element isotope separation program, the stable isotope separations program, and the radioisotope program were organized into one operating group known as the Isotopes Division. When the anticipated plutonium program was reviewed, it became evident that more plutonium isotopes were desired than were originally planned for. It seemed necessary to have available an ion source which could be serviced completely within a glove box and avoid the major expense of salvaging new calutron components after each run. It was also evident that such a design would have to be compatible with existing plutonium-containment equipment.

Fig. 1

Based on the above criteria the ion source shown in Fig. 1 was fabricated. The most radical deviation from past ion sources lay in the arc-chamber and charge-oven design, and the support structure shown in Fig. 2. The arc-chamber assembly is shown in the lower-left-hand of the figure. Ionization filament and electrodes are in position. In Fig. 3 the arc-chamber is in position and heater-buss connections are completed. The extent to which the ion source is dismantled for washing is shown in Fig. 4. In assembly to this point, all critical parts have been positioned by jigs. Those pieces which have been removed are machined to tolerances sufficient to provide self-alignment. Those components which comprise the arc-chamber and oven assembly are shown in Fig. 5. It will be noted that all parts are of graphite exclusive of charge container, heater insulators, and

Fig. 2

Fig. 3

external heater busses. Leakage of charge vapor from arc region into heater region is prevented by sealing rings shown in Fig. 6.

Fig. 4

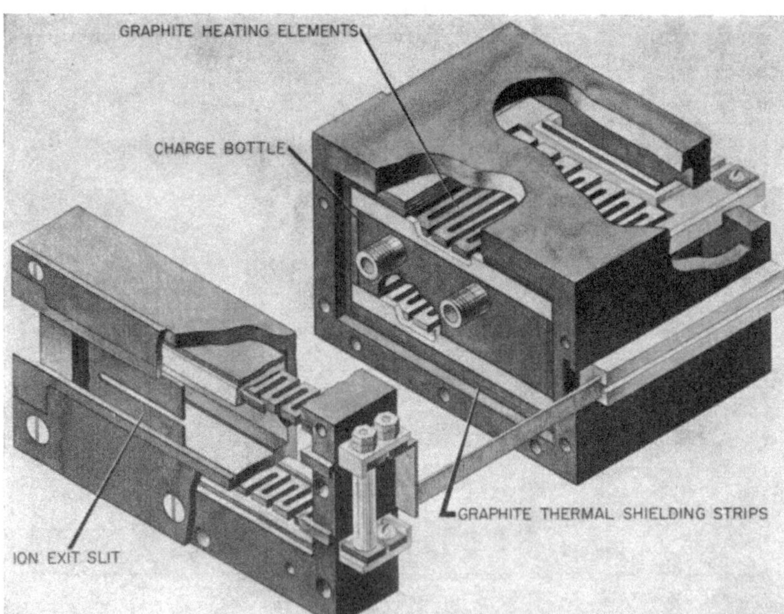
GRAPHITE HEATING ELEMENTS
CHARGE BOTTLE
GRAPHITE THERMAL SHIELDING STRIPS
ION EXIT SLIT

Fig. 5

This ion source has been used successfully with charges ranging from one gram of plutonium (mixed with 17 grams of uranium) to hundreds of grams of stable isotope charge feeds. Ion outputs have

Fig. 6

ranged from an average of 2 mA to 75 mA depending upon the ion exit slit dimensions and the element being processed. Typical values associated with plutonium separations are 1/16″ · 2″ ion exit slit, 2 mA received ion current, and 5% process efficiency obtained while limiting angular divergence of the received beam to 6 degrees.

Since this ion source was introduced into the plutonium separations, no ion source disposal has been made necessary by a mechanical failure. One metallic faceplate and ion source sub-assembly has been used in all runs performed since July 1957, and only the arc-chamber charge-oven assembly and electrodes have been renewed. It is anticipated that renewal of the basic ion source will be necessary only to accommodate chemical recycle of different isotopic mixtures of plutonium feeds.

Recently, sufficient plutonium feed has become available to make the simultaneous operation of two arcs in the same calutron desirable. Consequently an ion source of this type, but having two arc-chamber charge oven assemblies, has been constructed and is ready for testing, first with stable isotope feeds. This particular source will permit either simultaneous or concurrent utilization of the arcs.

"Single Isotope" Receiver Description

In radioisotope separations, more than in stable isotope work, it is frequently desirable to emphasize or to collect one nuclide. On this premise a collector was constructed to incorporate features which would reduce contamination and improve retention of the isotopic product. The receiver is shown in cutaway section in Fig. 7. It will

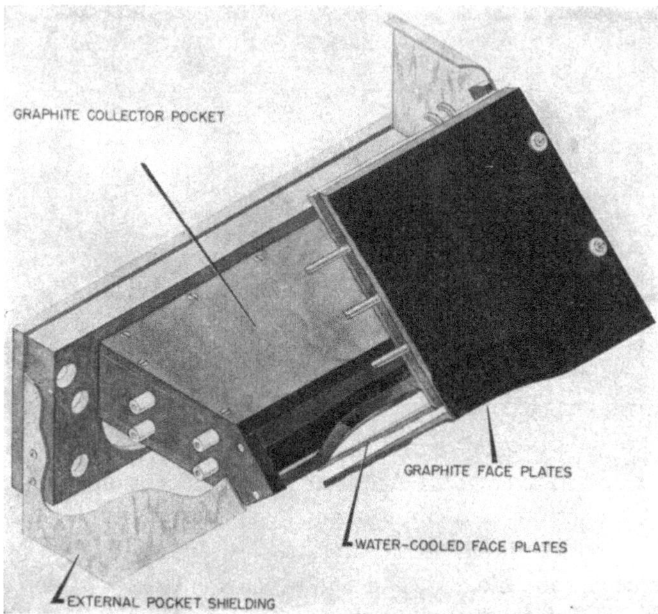

Fig. 7

be noted that the concept of dual beam defining slots has been used as first indicated by G. Ranc at the Harwell Conference in September 1955. Water-cooling or other means of refrigeration is incorporated only on the second set of defining plates. Another feature includes a provision for monitoring the current to each section of the first set of plates. Such provision is an invaluable aid in evaluating beam focus and total ion output. Pocket position is such that sputtering losses are minimized: the beam strikes obliquely after having spread out and the surface struck is opposite a well-cooled surface to provide for condensation of sputtered material. The contaminating effect of ion beams rejected by the first set of defining plates appears to be offset by other advantages as our best product has invariably been collected in this "single-isotope" arrangement.

A word of caution should be added concerning the rejection of the more abundant isotopes during use of this receiver. Beams are rejected at or near their best focal point, consequently, severe erosion

Fig. 8

of the defining plate is encountered. Evidence of this is presented in Fig. 8, which shows a .020-in wide channel cut 3/16-in deep in graphite by a 3 mA U^{238} ion beam. Operating time was approximately 30 hours.

The (Heavy Element) Separations Programs

The scope of the separations of uranium and plutonium is indicated in Table 1 which follows:

Table 1. *Summary of Special Separations since December, 1957*

Desired Isotope	Series Designation	Total Runs	Total Ion Reception Hours	Average U^+ or Pu^+ (mA)
U^{233}	J2S	8	781	2.1
U^{234}	B2S	4	72	1.8
	N2S	2	84	2.3
	Q2S	6	169	2.4
U^{235}	E2S	16	477	90.6
	F2S	5	223	2.6
U^{236}	A2S	4	160	2.2
	C2S	4	380	4.0
	H2S	7	172	17.5
U^{238}	G2S	36	1212	7.4
	K2S	3	167	103.4
	L2S	13	634	144.5
	O2S	15	829	149.7
Pu^{240}	D2S	2	97	0.3
Pu^{241}	I2S	3	73	2.0
	P2S	8	411	2.9

In each of these series of separations a specific approach was utilized to produce material meeting desired requirements. For instance high purity U^{233} was desired; and in series J2S, gram quantities were provided as indicated in Table 2.

Table 2. *Mass Analysis of* U^{233} *Samples*[1] *Series J 2 S*

Isotope	Mass Analysis (Atom %)			
	Feed	Sample		
		30 (a)	30 (b)	30 (c)
232	—	—	<0.0003	Incomplete
233	99.897	>99.9998	>99.9996	,,
234	.013	<0.00008	<0.0001	,,
235	.017	<0.00003	<0.0001	,,
236	—	<0.00003	<0.00005	,,
238	.073	<0.0001	<0.00005	,,
Sample Wt. (gm)		2.690	3.460	

U^{234} operations are limited by feed availability and repeated separations have been performed on chemically recycled calutron feed. Typical products achieved are indicated by results shown in Table 3.

Table 3. *Mass Analysis of* U^{234} *Sample Series N 2 S*

Isotope	Mass Analysis (Atom %)	
	Feed	Sample 46(a)
234	12.7	97.53
235	23.11	1.72
236	0.67	—
238	63.52	0.75
Sample Wt. (gm)		0.168

[1] Mass analyses of samples performed by Leo Collins, K.A.P.L.

The U^{235} separations have permitted both high purity and large quantity commitments to be fulfilled. The output from the same type of ion source was increased from 2.4 mA maintained in high purity

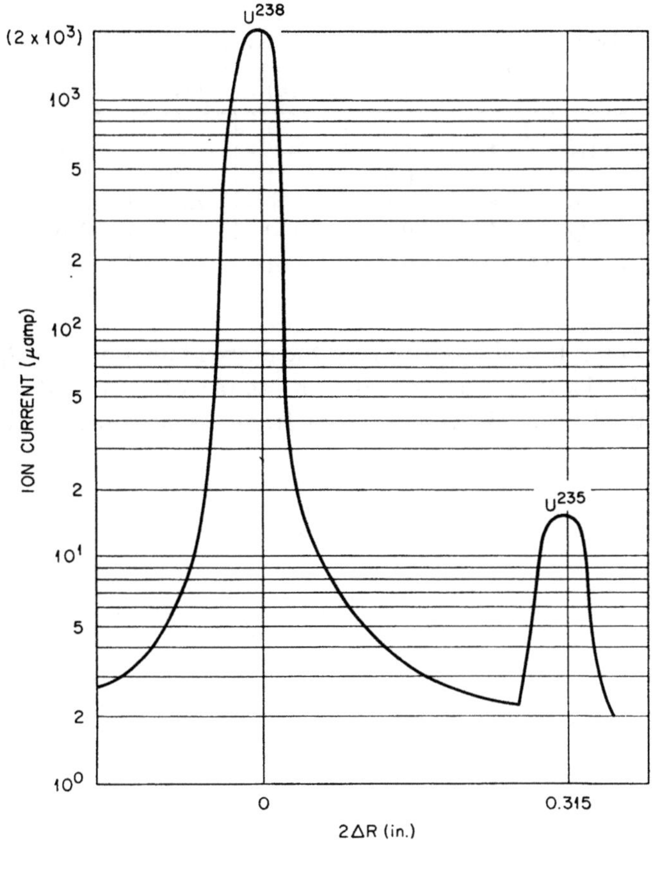

Fig. 9

work to 90 mA in the separation requiring a larger quantity of material. Development runs have enriched U^{235} from 0.7% to 93.9%. Beam resolution as shown in Fig. 9 was attained. Ion current was measured after passing through a 0.040-in beam defining slot.

The effectiveness of selectively sampling collected material is evidenced in the data for U^{236}, Series A2S. The analysis of four samples is shown in Table 4.

Table 4. *Mass Analysis of* U^{236} *Samples Series A 2 S, Runs 3–4*

Isotope	Mass Analysis (Atom %)				
	Feed	Sample			
		3 (A)	3 (B)	3 (C)	3 (D)
234	1.7	—	—	—	—
235	68.0	0.99	4.11	6.95	1.17
236	18.2	98.55	95.00	91.24	98.19
238	11.0	0.46	0.89	1.65	0.64
Sample Wt. (gm)		0.094	0.015	0.058	0.106

It is noted that the purity is in the descending order, A, B, C, D. The pocket location from which each sample was removed is shown in Fig. 10. Sample 3 (A) was taken from the area subjected to ion

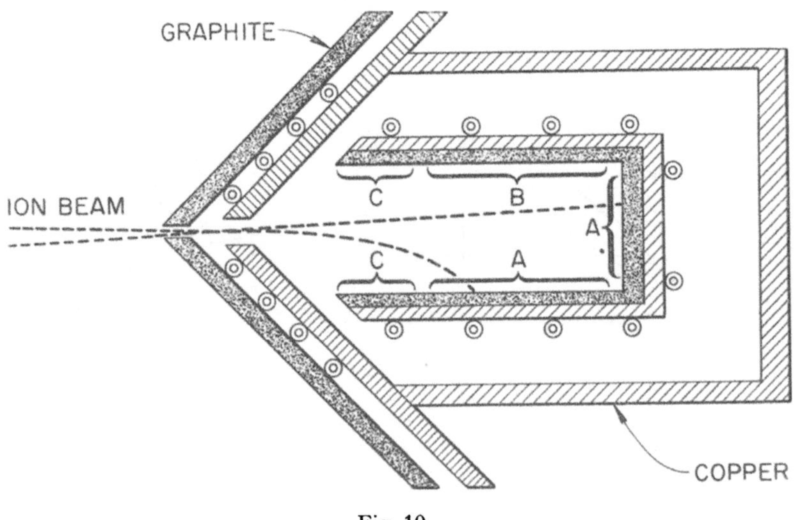

Fig. 10

impact; 3 (C) was the pocket area nearest the defining slot; 3 (D) was the two end areas of the pocket; and 3 (B) was the area opposite that on which the ion beam impinged.

U^{238} separations have also been performed to fulfill both high-purity and quantity requests. Outputs have averaged from 7.5 to 150 mA and much of the material contains no more than 7 ppm total isotopic contaminants.

Uranium separations are performed in six calutrons of 24-in radius located adjacent to the plutonium facility. Operation is regulated by demand for isotopic material and by chemical recycle time involved with the use of limited amounts of feeds. No major change in the uranium program is anticipated.

Plutonium separations are confined to a single 24-in radius calutron. Operating periods have been severely limited by the availability of feed materials. Current interest centers around gram quantities of Pu^{241} and recent runs have emphasized the collection of this isotope from less than 50 grams of plutonium containing 3.76% Pu^{241}. Chemical recycle of this feed required one week, unit handling and operation consumed another week; thus the plutonium P2S series was performed at the rate of approximately one run each two weeks.

Results obtained in the separation of Pu^{241} are shown in Table 5.

Table 5. *Mass Analysis of* Pu^{241} *Samples*

Isotope	Feed	Mass Analysis (Atom %)			
		Samples			
		29 (a)	29 (b)	29 (c)	55 (a)
239	66.42	1.10	1.29	2.67	7.74
240	29.04	2.18	2.09	6.15	10.42
241	3.76	96.60	96.44	90.64	81.31
242	0.78	0.12	0.18	0.54	0.53
Sample Wt. (gm)		0.022	0.005	0.021	0.319

In analyzing these results it should be understood that the plutonium facility shares the magnet yoke with a cyclotron. Certain magnetic field perturbations and voltage fluctuations result from cyclotron operations. Samples 29 (a) and 29 (b) were separated during cyclotron down-time; sample 29 (c) was separated partially during cyclotron operation; sample 55 (a) was separated while the cyclotron operated. Program commitments in both areas created this situation wherein both machines operated simultaneously.

At present the plutonium facility is not operating and modifications are being designed to improve total plutonium containment. In this concept a plutonium calutron and seven uranium calutrons are to be encased in an auxiliary room providing, at negative pressure, a passageway to the wash and servicing areas. This approach minimizes the possibility of contaminating a large area, if rupture of the plastic unit closures providing containment should occur. When this change is completed there is a commitment to separate a 10 gram sample of Pu^{241} within a year from 550 grams of plutonium containing 8.45% Pu^{241}. On the assumption that samples 29 (a) and 29 (b) are within a realistic operating range, the 10 gram sample should be enriched to between 95% and 98% Pu^{241}.

Distribution of Material in a Collector

Let us digress for a few minutes to talk about ion retention in general. For a long time difficulty has been recognized in predicting accurately the retention of different ions on different targets — or for that matter the same ion on the same target. Recent experience with a lead separation illustrates this point. Table 6 contains a summary of two operating experiences.

Table 6. *Retention of Lead on Graphite*

Operation	Isotope	Avg. Pb^+ (mA)	Metered Pb (gm)	Recovered Pb (gm)	Retention (%)
1	Pb^{206}	6.9	91.9	75.7	82.0
2	Pb^{208}	14.4	195.	3.5	1.8
	Pb^{204}	0.4	5.5	0.35	6.4

From this data it must be concluded that in the second operation the temperature within the receiver was elevated to the point that the lesser abundant Pb^{204} was lost along with the other lead isotopes.

In the past year several collectors have been constructed of dissimilar materials — that is the surface struck by the beam may be graphite while the opposing side of the pocket is water-cooled copper, etc. From this arrangement orders of magnitude measurements have been made on both the ionic and target material on the various surfaces. These data are included in Table 7 and in Table 8.

Table 7. *Retention of* Pb *on the Surface of* Pb^+ *Impact*

Ion	Target Material	Avg. Pb^+ (mA)	Metered Pb (gm)	Recovered Pb (gm)	Retention (%)
Pb^{204}	C	0.28	0.369	0.161	44
Pb^{204}	C	0.31	0.276	0.108	39
Pb^{207}	Cu	4.4	5.80	2.61	45

Table 8. *Weight Ratio of Target to* Pb *in Sputtered Material*

Ion	Target Material	Avg. Pb^+ (mA)	Pb Metered to Pocket (gm)	Metered Pb Sputtered (%)	Ratio Target/Pb
204	C	0.31	0.276	26	2.5
207	Cu	3.8	1.07	22	50.0

Each value shown is the result of a single determination. They are practical values expected to vary with time and possibly with beam intensity. Similar measurements with uranium are shown in

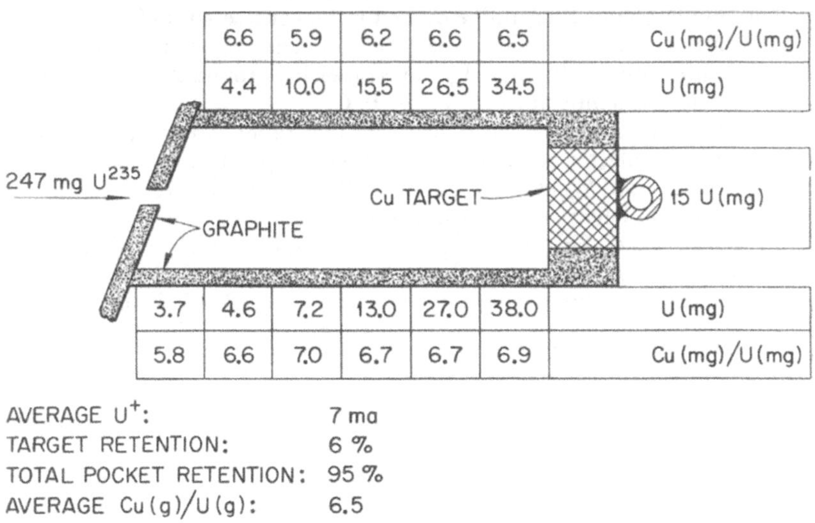

Fig. 11

Fig. 11. In this experiment a water-cooled target (½in × 2 in) was struck with a 7 mA, 35 kV ion beam. The metered amount of U^+ striking the target was 247 mg. Graphite plates placed 1 in apart formed the top and bottom of the pocket to permit determination of Cu and U in sputtered material. The values determined on ½in × 8in strips are as indicated. All these data indicate that much of the ionic species sputters and that pocket design should, whenever spacing permits, provide surfaces for trapping this material.

A renovated facility is just coming into operation to provide larger quantities of stable isotopes.

The building which contains the plutonium facility was initially equipped with two magnets, each containing 36 calutrons of 24-in radius. One magnet (or track) has now been authorized for use in stable isotope separations. In placing the magnet back into operation it has been divided into four smaller and magnetically isolated sections to permit simultaneous separations of four elements. In doing this the capacity of the facility has been reduced from 36 to 30 calutrons.

Discussion

R. Bernas: What is the temperature at the collector where the lead isotopes are impinging?

L. O. Love: I believe the average temperature to be 600–700° C. The material rejected from the collector was condensed on the receiver side plates and was not sufficiently enriched to be considered useful.

R. Bernas: We have found that the best temperature to condense lead vapour (pressure $\approx 10^{-3}$ mm Hg) was somewhat around 200–300° C. In no case should the temperature be below 0° C.

L. O. Love: Our experience has been that in all cases it is desirable to maintain the collector at the lowest possible temperature.

J. Kistemaker: What is the use of U^{238} enriched $10^9 : 1$.

L. O. Love: I do not know the reason.

J. Kistemaker: The sputtering ratio of U on Cu is 6.50, at what energy?

L. O. Love: The energy of the impinging ions was 35–38 kV.

P. K. Rol: The sputtering ratio of U on Cu at 30 kV is 6.5 g Cu/g U, this means the sputtering ratio is $\sim 26 \frac{\text{Cu at}}{\text{U ion}}$.

P. Weinzierl: I would like to know the current density achieved in your machine and what are the limits for this current density according to your experience.

L. O. Love: What the current density limits are from a calutron is a difficult question to answer. We conducted an experiment with two of our 2 amp. power supplies connected in parallel using a grid type ion source. Three amps of ion current were obtained at the collector using Lithium as the charge material. The beams were not very well resolved.

P. Weinzierl: What was the material and the size of the surface?

L. O. Love: Cu. This collector was 36 cm long and the total entrance slit width about 4 cm. The beam, however, was dissipated over a much larger area because the pocket was about 20 cm deep.

Recent Modifications to Hermes

By

R. W. McIlroy, J. H. Freeman, and **K. J. Hill**

Chemistry Division, A. E. R. E., Harwell

With 10 Figures

The general layout of Hermes including the electron impact source unit has been described in detail elsewhere [1] and this paper will be confined to recent modifications. Preliminary runs with lead and uranium showed that the design requirements (high resolution at 1 mA and 4% efficiency) could be met, and since plutonium isotopes were urgently required, active runs were commenced without further delay. At that time no other high resolution heavy element separation results were available for comparison, and the Hermes performance was certainly better than that then obtained in calutrons.

During continued plutonium separations it was found that, although the high enhancements could be maintained, theconditions for the best focus were extremely critical and difficult to maintain during the entire run (about 48 hours). Operating conditions could not readily be reproduced from run to run, the source efficiency varied from 0.5–8%, and the collection from 150 μA to 1.7 mA.

To discover the causes of these effects required a systematic study of source unit parameters which could not be carried out on Hermes itself since it is not sufficiently flexible or accessible, and the best policy for the machine at that time was to produce well enriched samples even though in small quantity. Recently the small twenty-four inch Harwell experimental separator has been fitted with a 40 kV source supply to make it comparable with Hermes, and it is now being used for a systematic and comprehensive study of the source unit. Although the investigation is far from complete it is already apparent that some of the difficulties experienced in operating Hermes can be attributed to perturbations of the source unit magnetic field, and also to the inadequate focusing of the ion beam by the extraction electrode system.

Source Unit Studies

(a) Perturbations of the source unit magnet field

Fig. 1 A shows a diagrammatic representation of the source unit under ideal conditions. Figs. 1 B–D show how the magnetic fields due to the auxiliary current supplies for heating both the filament and the arc chamber produce perturbations of the main source magnetic field and result in the deformation and displacement of the constrained arc. Normal methods of magnetic field plotting cannot be used to study these effects under operating conditions.

Fig. 2 shows a source unit which was modified to measure the arc deformations directly. The movable rack on the front of the source unit holds a carefully aligned polished copper plate at an adjustable distance directly below the filament. The modified source unit was inserted into the separator as for normal operation and an argon arc struck between the filament and the copper plate. The area of impact of the arc was defined by a sharp burn on the copper plate.

Subsequent measurement of the burn mark (Fig. 3) enabled the position of the arc to be measured within 0.1 mm.

Fig. 4 shows the calculated and measured displacements due to the filament current alone. Because of the difficulty of computing the exact field due to the complex shaped filament, too close an agreement cannot be expected.

Fig. 5 shows the additional displacements due to the arc chamber heater currents.

(The graphs have been corrected for a slight permanent inhomogeneity of the focusing magnet field.)

These results can be summed up by the statement that the distance from the extraction slit to the "edge" of the arc is closely dependent on all sources of magnetic field in the arc chamber. It was then decided to investigate the relationship between this distance and the extraction efficiency. Keeping *all operating conditions constant* the filament was moved mechanically and the efficiency measured in various positions, with the result shown in Fig. 6, which exhibits maxima in output at a certain distance from the extraction slit. This demonstrates the need to be exact in setting up the filament in relation to the expected operating conditions. Furthermore, any changes in the values of the filament current, the heater current or the source magnet field will result in a change of output and efficiency due

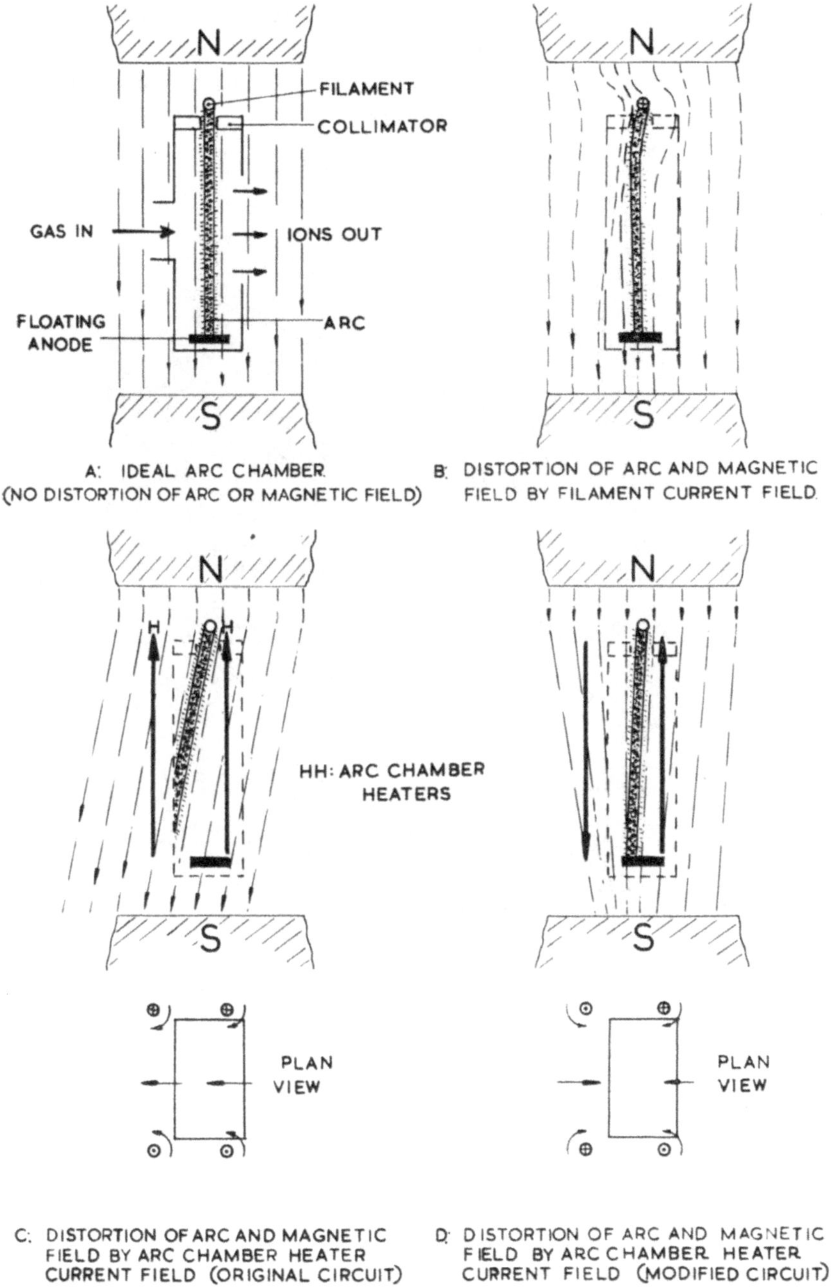
N
FILAMENT
COLLIMATOR
GAS IN
IONS OUT
FLOATING ANODE
ARC
S
A: IDEAL ARC CHAMBER (NO DISTORTION OF ARC OR MAGNETIC FIELD)
N
S
B: DISTORTION OF ARC AND MAGNETIC FIELD BY FILAMENT CURRENT FIELD
N
H
H
S
HH: ARC CHAMBER HEATERS
N
S
PLAN VIEW
PLAN VIEW
C: DISTORTION OF ARC AND MAGNETIC FIELD BY ARC CHAMBER HEATER CURRENT FIELD (ORIGINAL CIRCUIT)
D: DISTORTION OF ARC AND MAGNETIC FIELD BY ARC CHAMBER HEATER CURRENT FIELD (MODIFIED CIRCUIT)

Fig. 1

simply to displacements of the arc, and quite apart from the normal effect of these parameters on the arc. The filament may also bend the arc sufficiently to cause it to run into the narrow collimating slot which lies directly below the filament.

Fig. 2

In the forthcoming Hermes runs an attempt will be made to minimise these effects by

(i) widening the collimating slot,

(ii) offsetting the filament with respect to the collimating slot to allow for the induced curvature of the arc,

(iii) repositioning the filament and the collimating slot so that under normal operating conditions the arc lies in the position corresponding to maximum output and efficiency,

(iv) rewiring the arc chamber heaters so that the resulting field is minimised (Fig. 1 D).

(b) Extraction electrode geometry studies

During the preliminary studies on Hermes it was found [1] that a simple focusing plate at source potential provided a powerful control of the ion beam divergence in both planes. The size of the

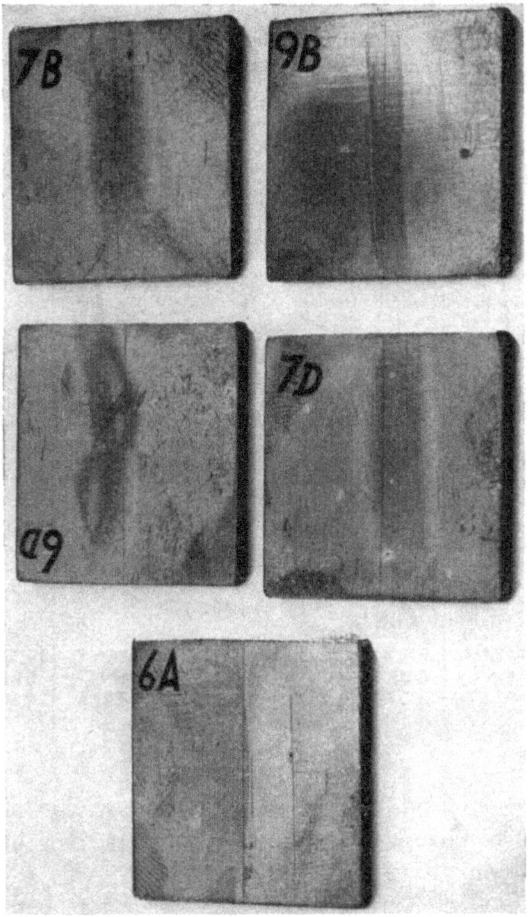

Fig. 3

hole in the focusing plate was adjusted to give a parallel beam in the direction perpendicular to the plane of the trajectory and a total divergence of about five degrees in the plane of the trajectory.

Experiments in the small separator have shown that with this geometry the undesirable lengthwise divergence is very sensitive to the beam intensity and that a parallel beam is only obtained over a very limited range of outputs. Since an angular divergence of only one degree is sufficient to increase the beam height at the collector unit by 8 cm (comparable to its total height), this effect will seriously restrict the transmission and the collection if the beam intensity is changed from its optimum value.

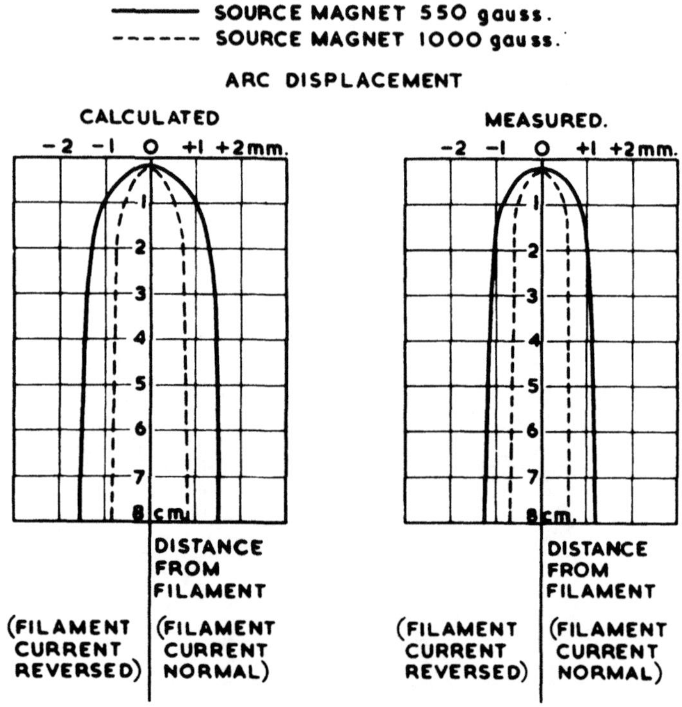

Fig. 4

It was also found that at constant output this divergence was independent of changes in the arc parameters, but that it was very dependent upon

(i) accelerating voltage,
(ii) electrode spacing,
(iii) size and shape of the focusing plate,
(iv) length of extraction slit.

Careful measurements were made for a series of geometries and Fig. 7 shows the modified geometry which was finally chosen. The spacing of the electrodes has been decreased, the slit lengthened and the focusing plate shaped to approximate to a Pierce electrode. Fig. 8

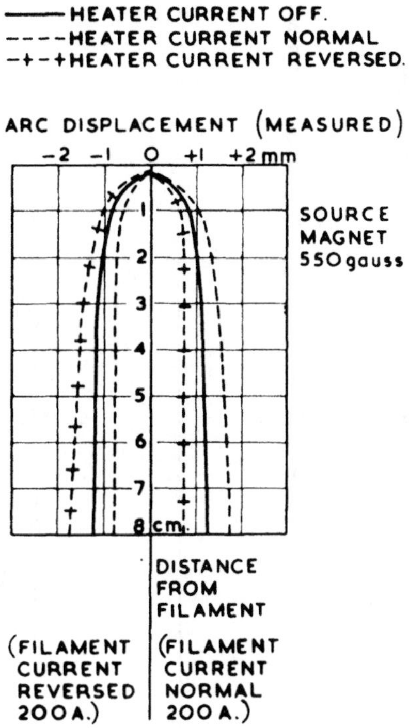
HEATER CURRENT OFF.
HEATER CURRENT NORMAL
HEATER CURRENT REVERSED.
ARC DISPLACEMENT (MEASURED)
-2 -1 0 +1 +2 mm
SOURCE MAGNET 550 gauss
1 2 3 4 5 6 7 8 cm.
DISTANCE FROM FILAMENT
(FILAMENT CURRENT REVERSED 200 A.)
(FILAMENT CURRENT NORMAL 200 A.)

Fig. 5

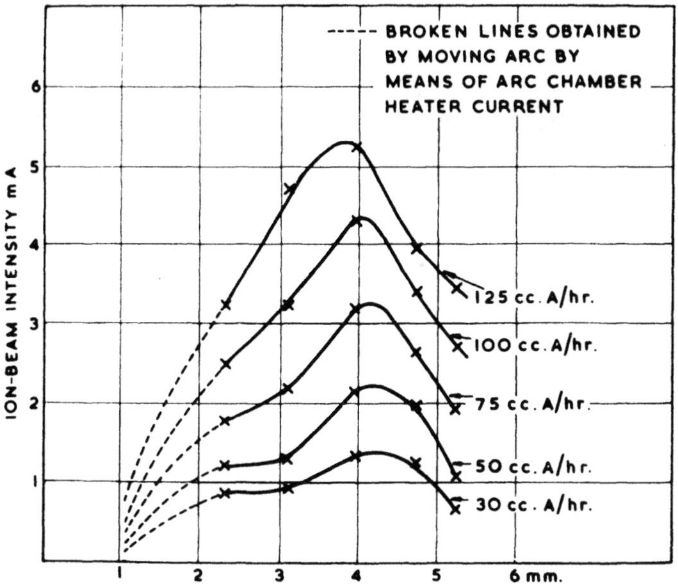
BROKEN LINES OBTAINED BY MOVING ARC BY MEANS OF ARC CHAMBER HEATER CURRENT
ION-BEAM INTENSITY mA
125 cc. A/hr.
100 cc. A/hr.
75 cc. A/hr.
50 cc. A/hr.
30 cc. A/hr.
1 2 3 4 5 6 mm.
DISTANCE OF FILAMENT CENTRE FROM EXTRACTION SLIT.

Fig. 6

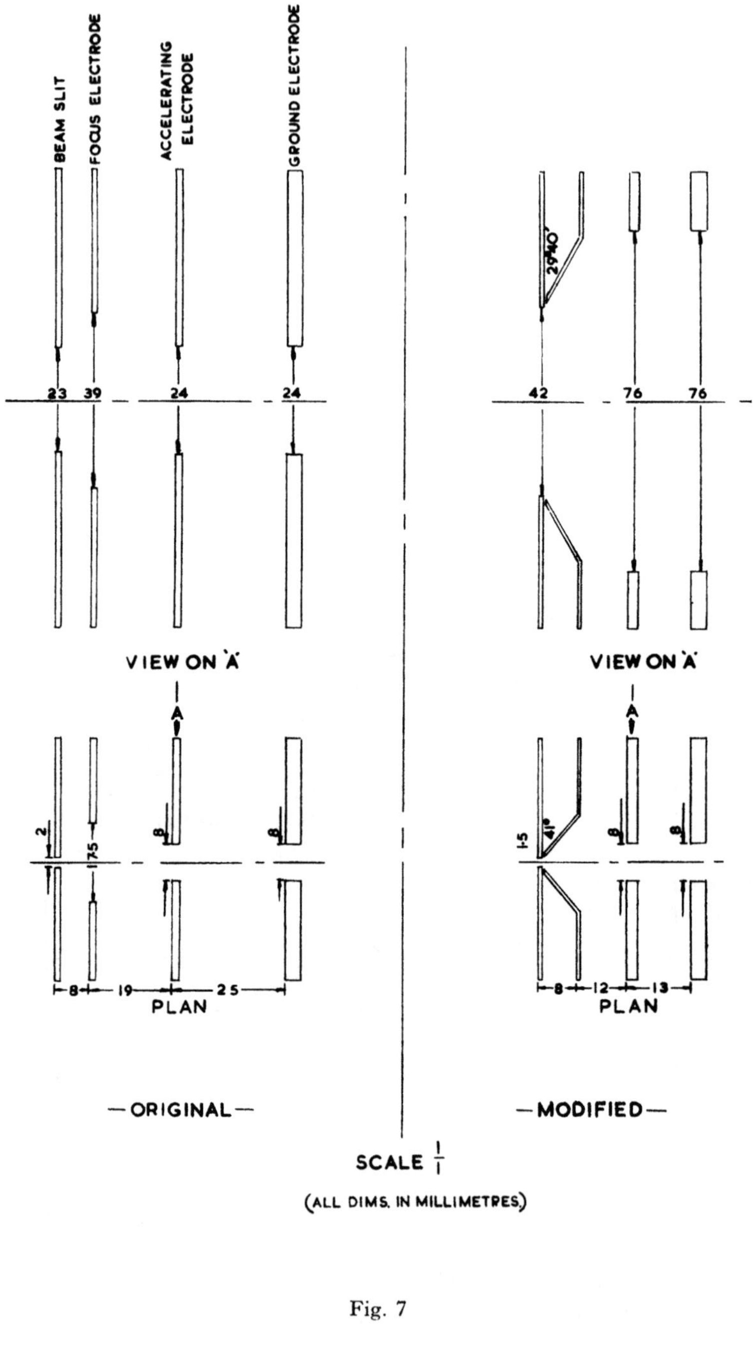
BEAM SLIT
FOCUS ELECTRODE
ACCELERATING ELECTRODE
GROUND ELECTRODE
23
39
24
24
29°40'
42
76
76
VIEW ON 'A'
VIEW ON 'A'
A
A
2
1·75
8
8
1·5
41°
8
8
8
19
25
8
12
13
PLAN
PLAN
—ORIGINAL—
—MODIFIED—
SCALE 1/1
(ALL DIMS. IN MILLIMETRES)

Fig. 7

shows the divergence as a function of output for the original and the modified geometries. Although the latter does not provide complete space charge compensation under all conditions it is quite satisfactory for the range of outputs envisaged for Hermes.

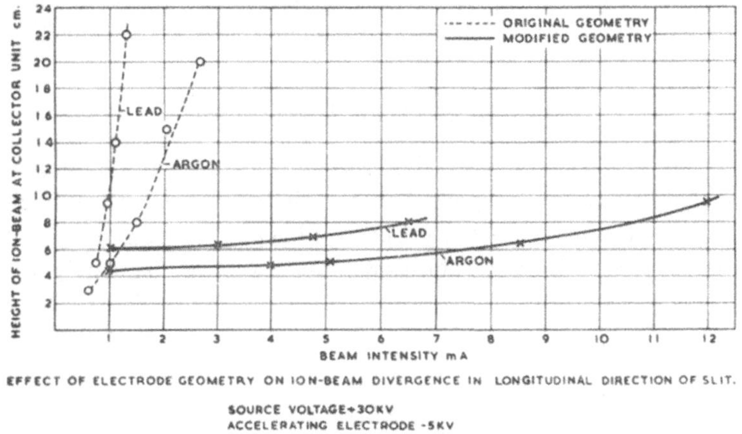

Fig. 8

Source Unit Movements

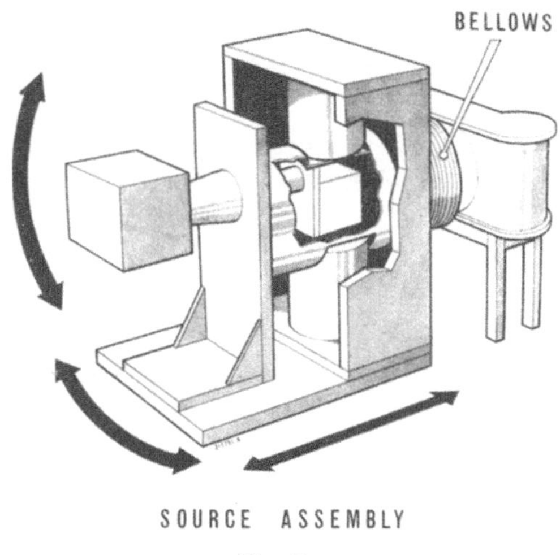

Fig. 9

With a 14 foot ion path a very small deflection from the true direction results in partial loss of the beam on the walls of the tank. This can arise from errors in setting up the source unit, or from variations in deflection due to the focusing magnet (which has the same effect on the beam as the main magnet but on a smaller scale). To make rapid adjustment possible the source unit has lately been connected to the main tank by a flexible bellows and the whole unit and focusing magnet together can be swung through arcs of 2° in the two planes at right angles which contain the initial line of flight of the beam (Fig. 9). By this means the beam can always be directed into the flight tank in the plane of the tank and tangential to the correct flight path. The horizontal swivel is arranged to work about the axis of the extraction slit which always remains in the correct position.

Collector Unit Movements

It has been found over the last two years that the focal point moves more than was expected in the directions along the final line of flight. This may be partly due to changes in the position of the virtual source with arc conditions, but can also be partly due to incomplete space charge compensation which gives rise to partial blow-up of the beam. If this is a small effect then it may result in a decreased rate of convergence and so the focal point moves back. The focus is presumably less good in this case but provided the effect is small it is convenient to accomodate it by moving the collector to the optimum position. A 12 inch travel is provided (Fig. 10).

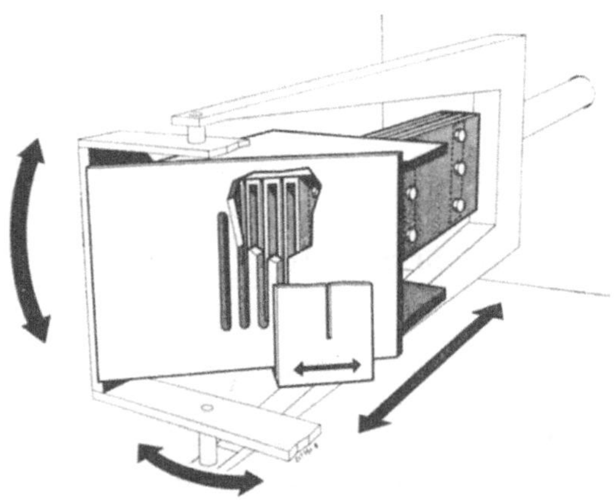

COLLECTOR UNIT

Fig. 10

It has also been found that sputtering losses are minimised if the beam strikes one side of the pocket at a low angle of incidence. The sputtered material deposits on the opposite wall of the pocket and stays there, provided that the beam is not subsequently swept over it. The collector assembly has therefore been given a swivel action about an axis perpendicular to the plane of the flight path which enables the beam to be played upon any part of the inside walls of the pockets. By fitting three insulated probes on the wall of the 239 pocket, at the front, middle and back respectively it is possible to set the collector assembly so that the beam is directed to one wall well inside the pocket. This reduces small angle sputtering from the front of the pocket and it also spreads the area of deposition and reduces the chance of localised overheating.

An additional movement about an axis parallel to the final flight path allows the slits and beam to be accurately lined up lengthwise without relying upon exact jigging. The less fine work required in pressure suits the better.

Reference

1. Smith, M. L., and K. J. Hill, Proc. International Symposium on Isotope Separation, p. 581, North-Holland (1958).

Some Characteristics of the E.M. Isotope Separator at the Nobel Institute of Physics, Stockholm

By

T. Alväger and **J. Uhler**

Nobel Institute of Physics, Stockholm

With 3 Figures

The electromagnetic isotope separator at the Nobel Institute of Physics has for a period of over ten years been used to a large extent in low energy nuclear physics research. During this period an extensive rebuilding of the whole machine has also been carried out; this was finished about two years ago [1].

We shall now give a short description of some details of the present performance of the isotope separator and a few characteristics of the instrument especially concerning the possibilities of obtaining very high purity of separated samples in a low intensity isotope separator containing only one separation stage.

Description of Machine

The separator is of the low intensity type which means that the total ion current is usually less than a few 100 μA while the efficiency measured as the ratio between the collected material and the amount of a certain isotope put into the ion source of the separator is, for easily separated substances, about 5–10%. The instrument is to a large extent used for the separation of radioactive isotopes of high specific activity obtained from cyclotron irradiation or from fission. The amount of activity put into the separator is usually weak, less than a few mC, but in a few cases stronger sources, of the order of 1 C, have been separated. In those cases, however, the half-lives of the activity have been short and no special radiation protection has been found necessary for the work of current interest.

The magnet has a mean radius of about 170 cm with 90° deflection. The source and collector are situated approximately symmetrically about the magnet. Due to this and to the large radius of the magnet the ion path length is relatively large, about 500 cm. From the focusing point of view this is, of course, not satisfactory. At small ion currents, however, the large ion path seems to be no drawback as it is possible to achieve very large separation factors in this case as discussed below. On the other hand the large dispersion obtainable — the distance between two masses at mass number $A = 250$, for example, is about 7 mm, while the half-width of the mass lines is usually less than 1.5 mm — makes the instrument especially suitable in many cases, as for example for investigations in the high mass region.

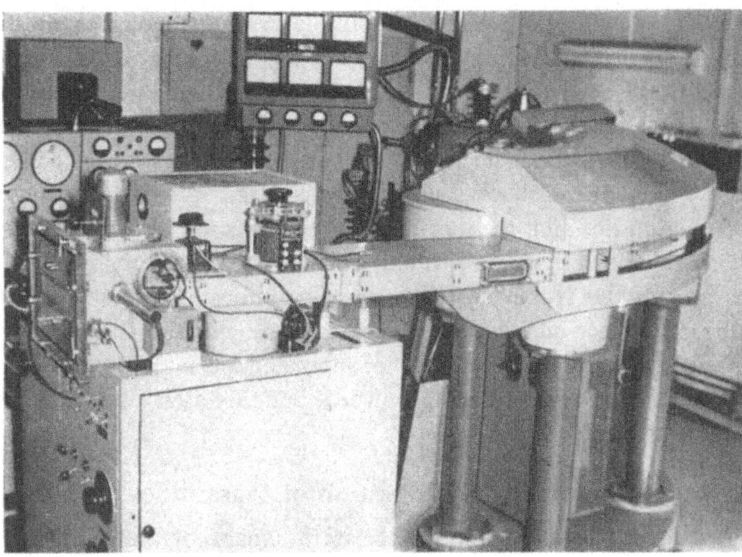

Fig. 1. View of the separator

The ion source is of the low voltage arc magnetic type. It is equipped with two ovens. This arrangement makes it possible to separate radioactivity in a very short time, which is of the order of a few minutes [2]. This is, of course, of special importance when the activity is short-lived (cf. [4]), but a rapid separation time is naturally always suitable as one can then, to a large degree, avoid instabilities and similar disturbing effects. Below an example will be given which shows the advantage of using the two-oven ion source in order to reach high resolution.

A more detailed discussion of the ion source is given in another paper in this symposium.

The acceleration system used at present consists of a four-cylinder electrode arrangement built together to one unit. This makes it possible to move the whole lens system during operation thereby varying the distance between the ion source and the first electrode without changing the distance between the other electrodes.

The optimum lens system which should be used in instruments of the type discussed here has not yet been discovered and several different designs have been made. One special question is how many electrodes should be used. The simplest system consists of three electrodes, the extraction electrode on earth potential, followed by an electrode on focusing voltage and an earthed one. Sometimes, however, a fourth electrode is used inserted between the ion source and the first earthed electrode and at a potential of the order of 10 kV below the potential of the ion source. No large systematic investigation to find the optimum conditions on the two types of lens systems has been performed, but a few experiments made at the separator of the Nobel Institute seem to indicate that the aberration of the four electrode system is less than that of the three electrode arrangement. This is especially pronounced at higher currents. As early as at 30 μA the resolving power is roughly a factor of two better in the four electrode lens system.

Intensity Distribution of Mass Lines

A property of large importance is the shape of the intensity distribution of a separated mass-line. Fig. 2 shows such a mass-line for a special case. This was obtained by collecting I^{131} for about 1 min. on an aluminium foil. This was then cut into 5 mm strips on which the activity was counted in a scintillation spectrometer. In the vicinity of the 131 mass-line the activity on 0.5 mm strips was counted. The activity obtained per 0.5 mm strip is given in the figure as a fraction of the peak-height intensity. The peak at mass number 132 is due to HI.

The intensity distribution can also be obtained by a special arrangement consisting of an oscillating probe in the beam at the collector from which the signals can be fed to, and registered on, an oscilloscope [1]. With this it is, however, usually possible to follow the intensity decrease only to a factor of about 100 below the peak intensity. To study those effects which especially contribute to the tails of the mass-lines, a more sensitive method is necessary, such as, for example, the technique shown in Fig. 2. This gives also the integrated distribution one really has in a separation, which the oscillating probe does not give.

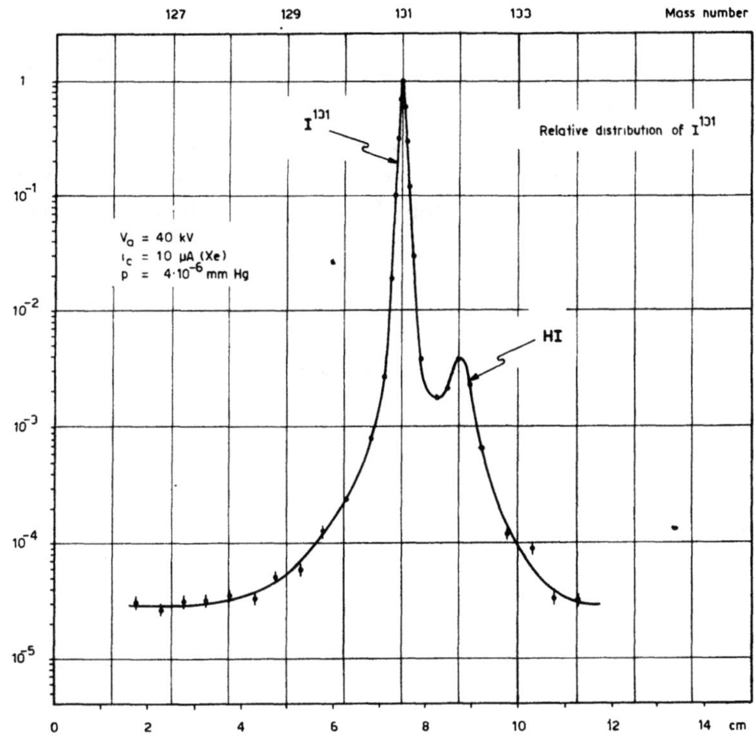

Fig. 2. Intensity distribution in semi-log scale of isotope-separated I^{131}

The curve in Fig. 2 is taken for accelerating voltage $= 40$ kV, ion current (Xe) $= 10\ \mu$A (used as mass-makers) and vacuum pressure $=$ $4 \cdot 10^{-6}$ mm Hg. At mass number $A = 130$ one finds from Fig. 2 $\mathrm{I} = 2.5 \cdot 10^{-4}$ pr 0.5 mm. For $A \leq 129$, I is about constant and equal to $3 \cdot 10^{-5}$.

The separation or enrichment factor F can be written for a mass A, when only one disturbing mass is present and if the distribution is taken to a first approximation as triangular, as:

$$F=\frac{C_f(1-C_o)}{C_o(1-C_f)}=\frac{1}{\mathrm{I}}\left(1-\frac{S}{4B}\right)\left(1-\frac{\delta}{4B}\right)^{-1}$$

(cf. [3]) where

C_o is the original concentration of mass A
C_f is the final concentration of mass A
I the itensity given in the figure
δ the width of the measuring slit ($=0.5$ mm)
S the width of the slit within which mass number A is collected
B the half width of the mass line

If we choose $S=2\delta=B$ which means that about 75% of the total intensity is collected within S, a value which can often be tolerated if high purity is needed, one finds for $A=130$ in the present case

$$F_{130}=3600$$

The maximum obtainable enrichment factor is found by use of the intensity at $A=129$. With this value one finds

$$F_{\max}=30,000$$

Preliminary measurements seem to indicate that the intensity in the region 5–15 mm from the peak intensity in the distribution curve can be reduced considerably by use of a simple shutter situated in front of the magnet. If the shutter decreases the beam intensity by about 5%, it seems to be possible to achieve the maximum enrichment factor even for mass number $A=132$ in the special case considered here. In most practical cases one has, of course, more than one disturbing mass. The separating factor decreases then but a value of 10,000–20,000 is quite realistic in the mass region discussed here. In the vicinity of $A=250$ a value of $F=5000$ is a more probable one.

As a comparison Fig. 3 will show the line shape of a mass line taken for a separation time of about 30 minutes. The activity is here Kr^{85}, and the distribution is measured in the same way as in Fig. 2. The separation data are, however, somewhat different from those used for I^{131}, but the curve seems to give the right tendency for longer separation times. It is seen that the tails are much wider here but they decrease to about the same minimum value as in the I^{131}-case. The larger tails in the Kr case should be due to small instabilities in the separator which are difficult to avoid.

The curve shows an asymmetry. This can partly be due to a small defocusing from the lens system, but a certain amount can be due to a capture process, where the ions in the magnetic field capture an electron resulting in a contribution to the high mass part of the line. Due to the presence of HI in the I^{131}-case, it is not possible to study this there. The capture part should in principle be avoidable by utilizing an electric deflection field placed behind the magnet. The ion beam could then be separated in a simple way from the neutral beam, which should give the tails at the high mass part of the line. This has, however, not yet been tried.

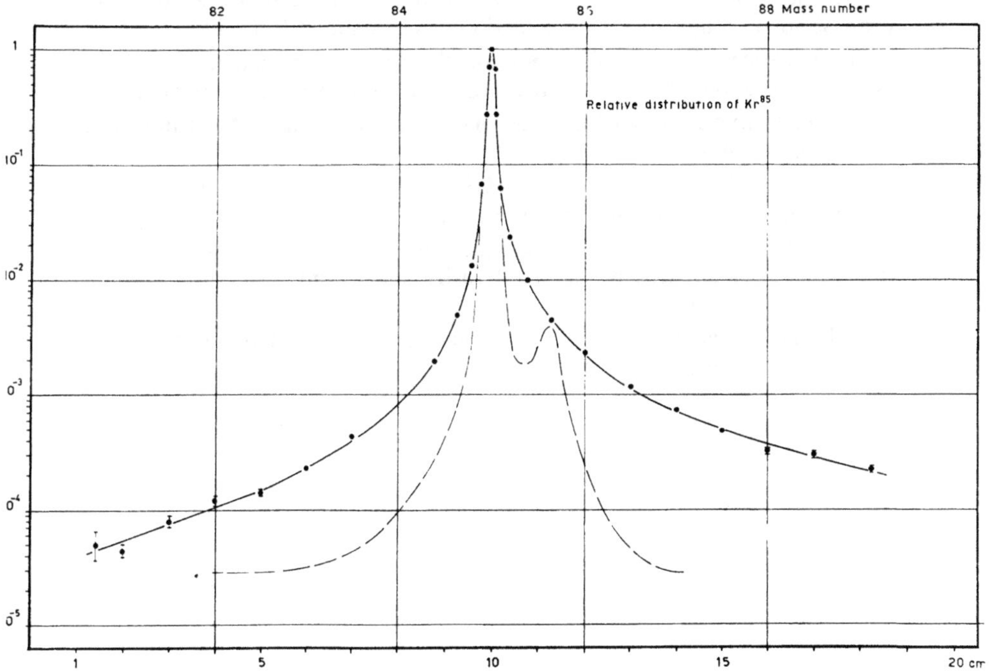

Fig. 3. Intensity distribution of separated Kr^{85} (30 minutes separation) and I^{131} (1 minute separation time)

It should be remarked that the separation time of 1 minute used in the I^{131} measurements is a normal separation time with the two-oven ion source. The conclusion from this is that by use of the short separation time method one can, when the ion current is relatively small, get a very large enrichment factor also by use of only one separation stage. The method is, however, only practical when the material used in the ion source is of high specific activity, but this is a very important field with a large number of applications.

References

1. ALVÄGER, T., and J. UHLER, Ark. Fys. **13**, 145 (1958).
2. UHLER, J., and T. ALVÄGER, Ark. Fys. **14**, 473 (1958).
3. ALVÄGER, T., Ark. Fys. **17**, 521 (1960).
4. ALVÄGER, T., Applications of E.M. separated radioactive isotopes at the Nobel Institute of Physics I p. 279.

Discussion

R. BERNAS: If the neighbouring mass were 100 times higher, what would the enrichment factor be? Did you not postulate that the two peaks were of equal height in your calculation of enrichment factor?

T. ALVÄGER: In the calculation of the enrichment factor no special assumptions about the relative magnitude of the two peaks were made.

CH. CASSIGNOL: Why is the experimental point of mass 84 not indicated in your figure No. 3.

J. UHLER: The missing point for Kr^{84} is due to the fact that the stabilizing arrangement by two pins shadows the target.

CH. CASSIGNQL: One could also make the following assumption: K^{84} is the most abundant isotope, it could easily sputter off some of the radioactivity deposited by K^{85}.

J. UHLER: The total current was very low $\sim 3\ \mu A$, I don't think that sputtering can be predominant.

Chapter IV

Progress in Ion Source Development

On Ion Sources with High Efficiency and Intensity

By

J. Kistemaker

F.O.M.-Laboratorium voor Massaspectrografie, Amsterdam

With 3 Figures

Summary

The conditions determining the shape of the ion emitting plasma cap in ion sources are analysed. Dense plasmas ($n_i = 10^{14}$) give expanding caps as in the Von Ardenne source. Thin plasmas ($n_i = 10^{10}$) give inward bended caps as in all normal ion sources. This difference in shape means a difference in ion emission with a factor ten.

The problem of an ion source with a high material efficiency is solved by the magnetic type ion sources. The Von Ardenne source has a high density plasma ($n_i = 10^{14}$) in the emission region with an extremely high electron temperature. Therefore hardly any neutral particles are available. In the magnetic P.I.G. type sources the material efficiency is obtained by internal ion focusing. They run with low plasma density ($n_i = 10^{10}$). For the separation of microgramme quantities of rare materials (radioactive or toxic like Pu) the P.I.G. type source should be preferred to the normally used arc source.

1. Discharge Boundary Conditions

To produce intense ion beams [1, 2], extracted from an exit opening at the end of the discharge chamber, it is necessary to create so many ions in front of this exit hole that a space charge limited current of positive ions can be withdrawn. The emission of the ions from the plasma boundary can take place either thermally, as in all low voltage arcs at sufficiently high pressures, or using a self focusing mechanism inside the discharge [10].

The shape of a plasma boundary is determined by the geometry of the extracting electrodes, the ion density of the plasma and the applied potentials. Suppose that the ion current emitted per unit area of this boundary is given by

$$J_b = n_i e (kT_e/m_i)^{1/2} \tag{1}$$

where n_i is the number of ions per cm^3 at the boundary inside the plasma, T_e is the electron temperature, m_i the mass of the ions and k the Boltzmann constant. On the other hand the space charge limited current from a flat emitter per unit area deduced via Poissons equation, for ideal vacuum, is [11]

$$J_{\text{sp. ch. l.}} = \frac{1}{9\pi} \sqrt{\frac{2e}{m_i}} \cdot \frac{1}{d^2} \cdot V^{3/2}$$

where d is the distance of the boundary to the extraction electrode, which has a potential difference V with the plasma boundary. The space charge limited current can only be withdrawn if J_b is high enough. The first requirement is therefore that

$$J_b \geq J_{\text{sp. ch. l.}} \tag{3}$$

which leads to a minimum ion density $n_{i_{\min}}$ opposite the exit hole of

$$n_{i_{\min}} = \frac{\sqrt{2}}{9\pi} \cdot \frac{1}{d^2} \cdot \{V^3/(ekT_e)\}^{1/2} \tag{4}$$

Example:

$V = 6 \cdot 10^4$ Volts $= 200$ e. s. u.; $d = 1$ cm;
$T_e = 10^4$, then $n_{i_{\min}} = 6 \cdot 10^{12}$ ions/cm^3 to give $J_{\text{sp. ch. l.}} = 800$ mA protons per cm^2 emitting area.

This is only true in the case of a flat plasma boundary and if no ions with a higher mass than the proton mass are present. If this $n_{i_{\min}}$ is not present the plasma boundary will be pressed backwards until the electrical field strength at the plasma boundary just fits with a space charge limited current. This is the reason why inward bended plasma caps arise in most ion sources where $n_i = 10^8$ or 10^{10}.

In the case that $n_{\text{plasma}} > n_{i_{\min}}$ the plasma will flow out of the source opening A (Fig. 1). The plasma behaves like a solid metal body if $n_{\text{plasma}} = n_i$ is large enough, and compresses the electrical equipotential surfaces ahead, actually increasing the field strength there. From

$$\frac{1}{4} e n_i v_i = \frac{1}{9\pi} \sqrt{\frac{2e}{m_i}} \frac{V^{3/2}}{d^2} \tag{5}$$

and $v_i = (2eV_{\text{plasmatherm}}/m_i)^{1/2}$, we find

$$d^2 \approx \frac{10^6}{n_i} \left(\frac{V^3}{V_{\text{pl.th.}}}\right)^{1/2}; \quad d \text{ in cm, } V \text{ and } V_{\text{pl.th.}} \text{ in Volts} \tag{6}$$

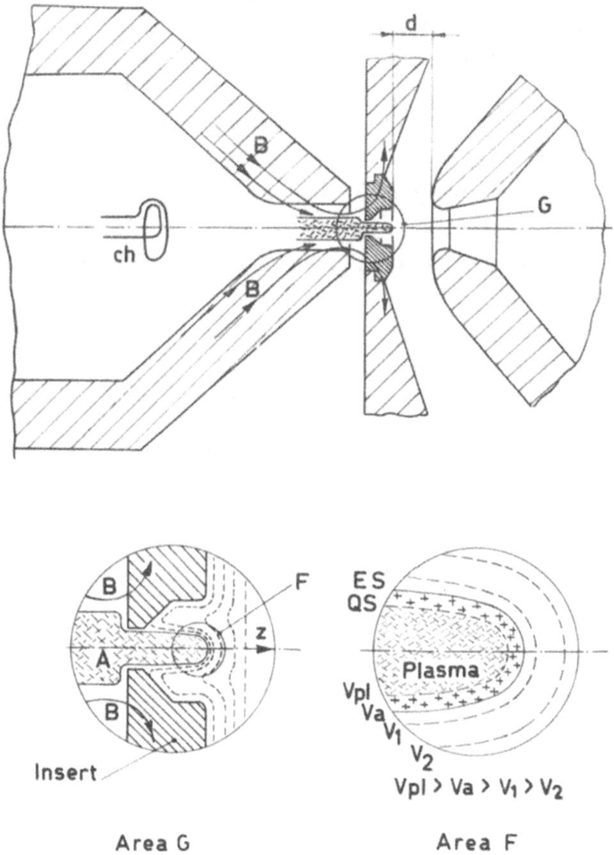

Fig. 1. Von Ardenne type duoplasmatron with protruding plasma cap
ch is the discharge chamber ES is the emitting surface at potential V_a QS is the plasma surface at potential V_{pl} The Insert is of Mo or W and has a very special shape to reduce the penetration of the extraction field

Table 1. *Relation between d, V_{plasma} and n_i*

n_i	$V_{\text{pl.}}$ Volts	V Volts	d cm	$E=\frac{V}{d}$ Volts/cm
10^{10}	0.1	100	0.5	200
10^{13}	10.0	100	0.006	$2 \cdot 10^4$
10^{14}	0.1	100	0.006	$2 \cdot 10^4$
10^{14}	300.0	100	0.0008	$1.2 \cdot 10^5$

We see from Table 1 that to keep up with the flow of positive ions from a plasma with density $n_i = 10^{14}\ \text{cm}^{-3}$, and with an apparent mean particle energy inside the plasma of 300 electron Volts, an electrical field strength E, short ahead of the plasma boundary, of 120 kV/cm is necessary. This field strength is not available in the insert of Mo (see Fig. 1). The penetration of the extraction field was reduced on purpose to give the plasma the opportunity to flow out. The outflow of the plasma is only stopped at that surface at which a space charge limited current just can be taken off, and equation (5) is satisfied.

It will also be clear now that working with a too low accelerating voltage can have disastrous consequences, if the value of n_i in the source plasma is high. The plasma expansion goes too far and one gets a very diverging ion beam, with big losses at the electrodes.

The value of v_i in eq. (5) can be increased by the force exerted on the plasma because of the decreasing magnetic field B, corresponding with

$$K = -\frac{e V_{\text{pl.th.}}}{B} \frac{\partial B}{\partial z} \quad \text{(per particle).}$$

In a strong magnetic lens this can give rise to an apparent doubling of the "thermal" speed with which the positive ions pass the surface QS (the plasma boundary).

The above considerations try to give an explanation of the very large currents/cm² at the exit area, which have been obtained with VON ARDENNE type ion sources (100 Amps/cm², protons). The explanation is mainly that the emitting area is easily a factor 10 larger than the small exit area A, because of the protrusion. The real shape of the protruding plasma inside and just outside the insert is not yet clear, however [2, 3, 13, 14].

The expansion of the plasma from the arc sources in large electromagnetic isotope separators (calutrons) is clear from cases 1, 2 and 3, Table 1, and also from [4]. In the normal case of $n_i = 10^{10}$ the plasma is bended inward ($E = 200$ V/cm). However, if for some reason (pressure rise, or presence of easily ionisable atoms like Li) the n_i increases, the plasma comes out at the exit slit and requires some $E = 10$ kV/cm to stop it. This can give rise to bad focusing, and therefore to bad efficiency.

2. Concerning Ion Source Efficiency

If one studies the ratio of the quantity of ions leaving an ion source to the total quantity of gas consumed by the source (the gas efficiency γ), one finds for various sources practical data indicated in Table 2.

Table 2. *Gas efficiency* γ

Type of ion source	Experimental gas efficiency	Operating pressure in source
Arc discharge	2–10%	10^{-3}–$3 \cdot 10^{-2}$ mm Hg
High frequency	5–10	10^{-2}–$3 \cdot 10^{-1}$
Magnetic (P. I. G.)	10–50	10^{-5}–10^{-2}
Von Ardenne	up to 100	$\sim 10^{-2}$

Arc discharge ion sources are used in the large electromagnetic isotope separators in Oak Ridge, Harwell, and Amsterdam [4, 6, 9].

High frequency sources are only used on particle accelerators. The Thonemann type is famous. Magnetic ion sources have been developed by ourselves (Kistemaker and Douwes Dekker) and by Nielsen and by many other people [10, 5, 7, 8, 12].

One can try to get high efficiency sources either by:
ionising the gas at the plasma boundary to a high degree or
using as low a pressure as possible in the source.

In both cases there is an extra requirement that the source body as a whole should be at high temperature to prevent recombination, absorption and condensation inside the source body. Moreover, trivial losses should be reduced, as for instance chemical decomposition of the charge material and bad focusing of the ion beam after leaving the source.

We will treat here only the main aspects of the ionisation of the gas, and a high gas economy (magnetic source).

a) High degree of ionisation

If the plasma at the boundary layer contains $n_i = n_e = n_{pl}$ charged particles per cm^3, and n_{gas} neutral particles the gas efficiency γ is determined by

$$\gamma = 100 n_i v_i / (n_i v_i + n_{gas} v_{gas})$$

Take an *example*:

$$\text{degree of ionisation} = \frac{100 n_i}{n_i + n_{gas}} = 50\% \text{ in } H_2 \text{ gas}$$

v_{gas} corresponds with 0.03 electron Volts (eV_{gas})

v_i corresponds with 30 electron Volts (eV_i)

$$\gamma = 100 \sqrt{2}\, n_i \sqrt{V_i} / (\sqrt{2}\, n_i \sqrt{V_i} + n_{gas} \sqrt{V_{gas}}) \tag{7}$$

We took $n_i = n_{gas}$, which gives:

$$\gamma = 100 \sqrt{2} \sqrt{V_i} / (\sqrt{2} \sqrt{V_i} + \sqrt{V_{gas}})$$
$$= 98\% \text{ in this case.}$$

Conclusions:

1. The degree of ionisation should be 10% or higher.
2. At the plasma boundary the plasma must flow out at a speed that is high in comparison with the mean thermal velocity of the gas particles.

To get the required degree of ionisation it is necessary that

$$\lambda_{ionisation} = (n_e \sigma_{ionisation})^{-1} < l \tag{8}$$

where $\lambda_{ionisation}$ and $\sigma_{ionisation}$ are respectively the free path length and the cross section of a neutral gas particle in the plasma with an electron density n_e and dimension l.

If $\sigma_{ionisation} = 10^{-15}$ cm^2, we see that

$$n_e l > 10^{15}$$

which shows why most gas discharges of low density have very low degrees of ionisation. Indeed l is always of the order of 1 cm, but electron densities of $10^{15}/cm^3$ are never achieved if normal arc currents are drawn. Such densities are only there if gas with an original pressure of 0.03 mm Hg is fully ionized. And the only way to keep such a gas highly ionized is to have an electron temperature $T_e > 10^5\,°K$, which corresponds to 10 eV_e.

Such extremely high temperatures have been reached in VON ARDENNE's [2] ion source, working at gas pressures more than 10^{-2} mm Hg in the following way:

1. *Large potential differences* on both sides of the plasma region between *A* and the boundary (see Fig. 1). Most electrons are accelerated over about 100 Volts after their liberation.

2. *Magnetic focusing* of the primary as well as the secondary electrons, causing a long lifetime and an excellent energy distribution over the electron sea inside the plasma via many electron-electron collisions.

b) High gas efficiency because of internal focusing [10, 12]

In ion sources with an axial magnetic field of the PENNING type (P.I.G.), as we have dėveloped in Amsterdam, operating pressures of 10^{-3} to 10^{-5} mm Hg have been used. According to eq. (8) this means that $\lambda_{\text{ionisation}}$ is between 10 and 1000 cm and in normal gas discharges this would mean a degree of ionisation lower than a few per cent. Our measurements have shown, however, that the axial magnetic field, combined with the electron reflection electrode *F* (see Fig. 2), build up a negative space charge in the middle of the

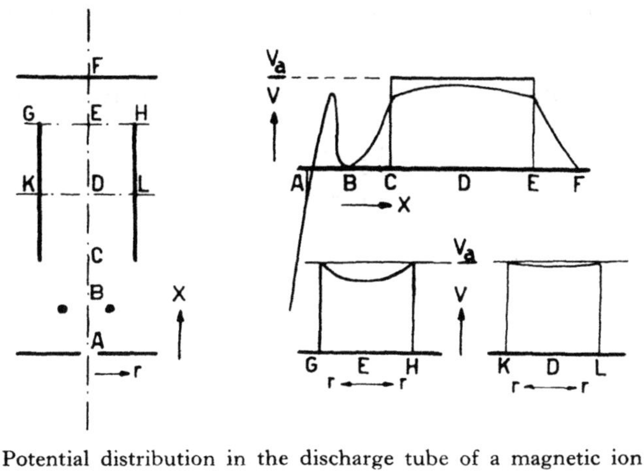

Fig. 2. Potential distribution in the discharge tube of a magnetic ion source

discharge tube. This negative trough makes it impossible for the positive ions, created in this tube, to reach the anode wall. They are all focused in the middle and fall from the potential mountain with the highest top in *D* (Fig. 2), guided by the axial magnetic field, to the exit opening A, where they have reached a speed corresponding to approximately $e(V_{\text{anode}} - V_{\text{cath}})$ which should be about 100 eV.

Half of the ions flow in the wrong direction and are lost and neutralized at the end of the discharge chamber. If the chamber is hot they can evaporate again and be used again and again, until they escape via the exit hole A either in ionic or gaseous shape.

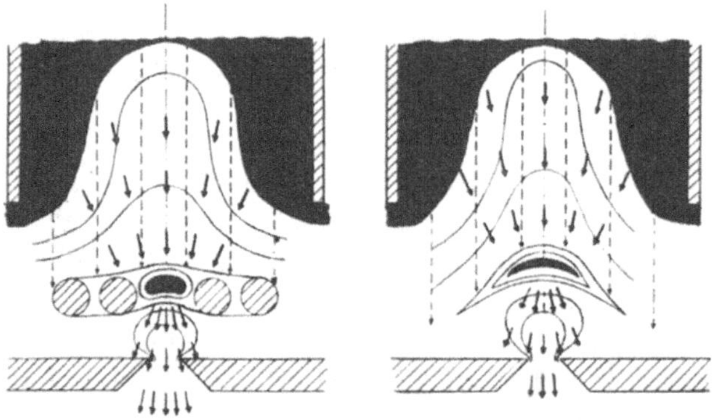

Fig. 3. Ion focusing inside the magnetic discharge chamber on the exit hole. Where the electrons are reflected a positive potential mountain might arise, sifting out low energy ions. These sources have an amazing small energy spread (less than 10 eV)

The optimum gas efficiency of this source is determined by

$$\gamma = \frac{100\,(n_i v_i)_{\text{sp. ch. lim.}}}{(n_i v_i)_{\text{sp. ch. lim.}} + \frac{1}{4} n_{\text{gas}} v_{\text{gas}}}$$

$$\text{where } (n_i v_i)_{\text{sp. ch. lim.}} = \frac{O}{9\pi} \sqrt{\frac{2}{e m_i}}\, \frac{V^{3/2}}{d^2} \qquad \text{[see eq. (5)]}$$

The factor 1/4 originates from the fact that all the ions have an axial direction and therefore no ion diffusion takes place, contrary to the gas.

The flow of ions inside the discharge to the orifice at A (Fig. 2) is determined by

$$n_{\text{gas}} n_e \sigma_{\text{ionis.}} \bar{v}_e \cdot \text{Volume}$$

If the space charge limited current through the orifice in area O is not reached the gas efficiency decreases. An increasing γ value can be obtained then by increasing n_e or by improving the magnetic focusing at the exit hole A. Increasing n_e is usually done by increasing the arc current. Very high material efficiencies have been obtained with these ion sources in Copenhagen by SKILBREID, and SIDENIUS [15].

Conclusions

Today there are ion sources with very high material efficiencies *viz.* the VON ARDENNE type and the PENNING type (P.I.G.). Both can have efficiencies of 50% or perhaps even higher. Both use an axial magnetic field in the direction of the ion extraction. The big difference between these two types is the shape of the plasma cap.

The P.I.G. ion source can produce ion currents (H^+) up to about 1 Amp/cm^2.

The VON ARDENNE ion source reaches 100 Amp/cm^2. For the separation of microgramme quantities of rare material (radioactive) the P.I.G. ion source is more elaborate and the extracted currents are sufficient. The VON ARDENNE source seems especially of importance for the separation of bulky quantities of isotopes and is certainly of importance for particle accelerators and plasma physics. Because of the very favourable material efficiency the pumping of the acceleration chamber is seriously reduced.

References

1. LAMB, W. A. S., and E. J. LOFGREN, R.S.I. **27**, 907 (1956).
2. VON ARDENNE, M., Tabellen der Elektronenphysik, Ionenphysik und Übermikroskopie I, p. 544–549 (1956).
3. MOAK, C. D., H. REESE, and W. M. GOOD, Nucleonics **9**, 18 (1951).
4. KISTEMAKER, J., P. K. ROL, J. SCHUTTEN, and C. DE VRIES, Z. Naturf. **10 a**, 850 (1955).
5. SOMMERIA-KLEIN, J., Ann. Phys., Paris, **13**, 344 (1956).
6. DRUAUX, J., and R. BERNAS, Proc. Harwell Conf. 1955.
7. ALMÉN, O., and K. O. NIELSEN, Proc. Harwell Conf. 1955.
8. BARNETT, C. F., P. M. STIER, and G. E. EVANS, R.S.I. **24**, 394 (1953).
9. KAMKE, D., Handbuch der Physik 33 (1956).
10. KISTEMAKER, J., and H. L. DOUWES DEKKER, Physica **16**, 198, 209 (1950).
11. LANGMUIR, I., and K. T. COMPTON, R.M.P. **3**, 237 (1931).
12. ALMÉN, O., and K. O. NIELSEN, Nuclear Instruments **1**, 302 (1957).
13. FRÖHLICH, H., Nukleonik **1**, 184 (1958).
14. MOAK, C. D., c. s., R.S.I. **30**, 694 (1959).
15. SKILBREID, O., SIDENIUS, G., (present volume) p. 243.

Discussion

W. Walcher: Dr. Kistemaker has reported values of the efficiency of 1%. I should like to recall to our minds the very fine experiments of Dr. Nielsen with a magnetically confined ion source which have shown that it is possible to obtain source efficiencies up to 25% and overall efficiencies up to about 10% without hash-disturbance.

Furthermore Dr. Kistemaker has stated that the plasma comes out of the exit hole in the shape of a pencil. It is my opinion that only from a well-formed meniscus it is possible to obtain a well defined beam and thereby high resolving power. From the pencil-shaped plasma boundary one can by no electron optical means form a narrow beam. The pencil which Dr. Kistemaker sees before the exit hole is—in my opinion—not the discharge plasma but some secondary phenomena.

J. Kistemaker: The Nielsen type of source is magnetically confined and has the same efficiency as the gas ion source which is indicated with 10–50% in Table 2.

W. Walcher: Who has any experience with the Von Ardenne type of ion source unoplasmatron or duoplasmatron? There do not seem to exist publications beside the "Tabellen zur Elektronen- und Ionenphysik und Übermikroskopie".

R. Bernas: I believe there are results published by Moak [1] and co-workers on a duoplasmatron source for a Van de Graaff in Oak Ridge.

J. H. Freeman: How can the duoplasmatron be used for high temperature solids?

R. Bernas: By cooling the iron pole pieces and introducing several heat shields between the discharge and the iron envelope.

G. Andersson: In the duoplasmatron ion source for solid material do you expose the cooled iron yoke to the discharge?

R. Bernas: No, of course not. We put 5 radiation shields of 0.1 mm thickness inside the discharge chamber and so we have more than 1000° C inside and water temperature outside.

K. O. Nielsen: Dr. Kistemaker mentioned that in order to get high efficiency it is necessary to have oscillating electrons. We only increased the heating current of the filament.

J. Kistemaker: Of course, if you increase the filament temperature you can always produce so many ions because there are more electrons emitted in comparison to the oscillating type.

Ch. Cassignol: I think that Prof. Kistemaker's analysis is far from being sufficient, since he does not consider at all the hash manifestations which are of very great importance in practice. Besides the trivial manifestations of instability, starvation hash and insulating layers, there are other ones, caused by the presence of a magnetic field, which have not yet been physically explained in a satisfactory manner but which govern the well known compromise between resolution and intensity and the bursting of the beams when the pressure inside the main tank (not inside the source) is lowered. A complete theory of ion sources would be very difficult to work out and physically and mathematically very complex. When the facts just mentioned are neglected, the theory can be worked out in an elegant and simple way but it has little reality.

K. O. NIELSEN: The ion current from an ion source is almost independent of the pressure in the chamber. Instabilities in ion sources take place at low pressure and, therefore, it is not surprising to find high efficiency when the source is hashy.

CH. CASSIGNOL: When our sources are hashy and the efficiency rises, it is not due, to a lowering of the pressure inside the source but to the value of the source magnetic field. As an average, in the Saclay separator, one can have steady beams when the applied field is about 50 Gauss. When this field is raised to 100 G and the other parameters are kept constant the beams are hashy and more intense but unresolved.

Reference

1. MOAK and al., R. S.I, **30**, 694 (1959).

An Ion Source for Short-Lived Radioactive Isotopes and Preliminary Results on a Duoplasmatron Type Ion Source

By

J. L. Sarrouy and **R. Klapisch**

Université de Paris, Laboratoire de Physique Nucléaire, Orsay (S. et O.)

With 6 Figures

Summary

This paper deals essentially with the ion sources which we are using at present or which we shall use in a short time on the double magnetic deflection separator of Orsay for work on radioactive isotopes.

We shall discuss the subject under three heads:

1. Description of the ion source for the separation of radioactive mercury obtained by the bombardment of gold in the synchrocyclotron.
2. Range of short-lived radio isotopes which may be processed in this source.
3. Preliminary results on a duoplasmatron type ion source.

The separation of short-lived isotopes must take account of the following two facts:

a) the amount of radioactive material is generally extremely small and highly diluted in the original element,

b) the efficiency of the separation is, in the present state of matters, very small.

Starting from this, one must admit that the time elapsing between the end of the irradiation and the beginning of the study of the deposit obtained at the collector of the separator must be very short.

For the radioactive mercury isotopes which are the subject of this first part we have succeeded in reducing to 3 minutes the time needed for the separation by making use of one of our usual sources to which we have added a quick working vacuum lock. Mass identification of the mercury isotopes was done by evaporating in the source some compound capable of giving intense ion beams in the mass region used.

In this particular case a mixture of rare earth chlorides with a high proportion of $GdCl_3$ and some $DyCl_3$ has given, among others, ions of $GdCl^+$ and $DyCl^+$ which act as excellent markers in the region of masses 180 to 200.

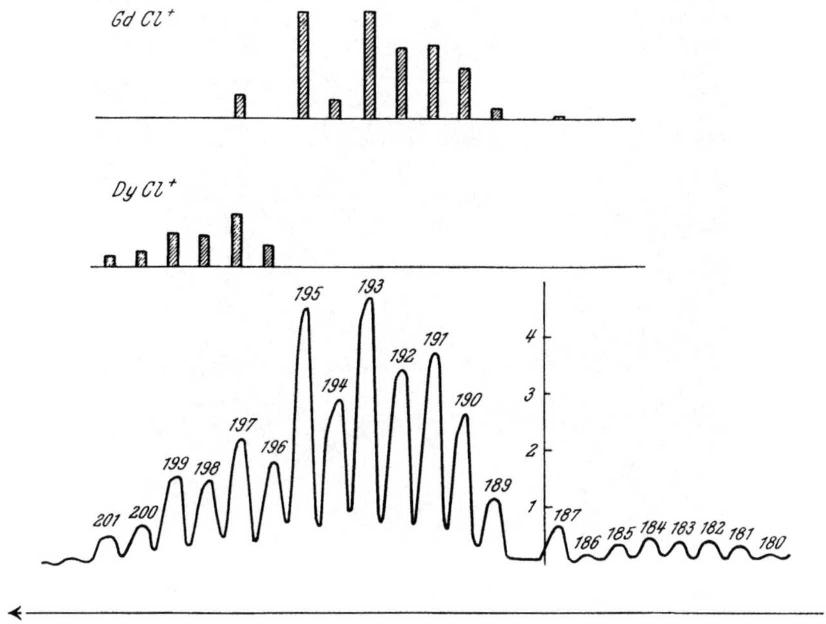

Fig. 1. Theoretical spectra of $GdCl^+$ and $DyCl^+$ ions. Comparison with mass spectrum record of ion beams coming from the source

The comparison between the theoretical spectra and the recordings obtained at the collector are shown on Fig. 1; after a few separations it is easy to recognize, in the pattern visible through the collector window, what masses we are dealing with.

Fig. 2 is a photograph of the ion impacts on the collector in this region of masses.

Another advantage of the ionised chlorides is that the discharge in the source is practically not perturbed when the small sample holder containing the radioactive sample is introduced. The relatively high temperature of this source permits a quick and total extraction of the mercury contained in the gold foils.

Fig. 2. Photograph of impacts on the first stage collector of the $GdCl^+$ and $DyCl^+$ ion beams

Some numerical values of the various parameters of the source are given in Table 1 below.

Table 1

Heating of cathode	48 A — 5 V (240 W)
Arc	1.5 A — 200 V (300 W)
Heating of oven	5 A — 5 V (25 W)
Total power in the source	500 to 600 W
Approximate temperature of $GdCl_3$	700° C
Approximate temperature of the ionisation chamber	800 to 1000° C

The sample holder is a cylindrical tube made of stainless steel 5 mm in diameter and about 10 mm long; it weighs approximately 0.4 gr.

Fig. 3 shows a section of the ion source and of the ion optics.

After the source has operated for a certain time, long enough to become stable, the sample holder is introduced in a lock of a few cubic-centimeters shown in the figure.

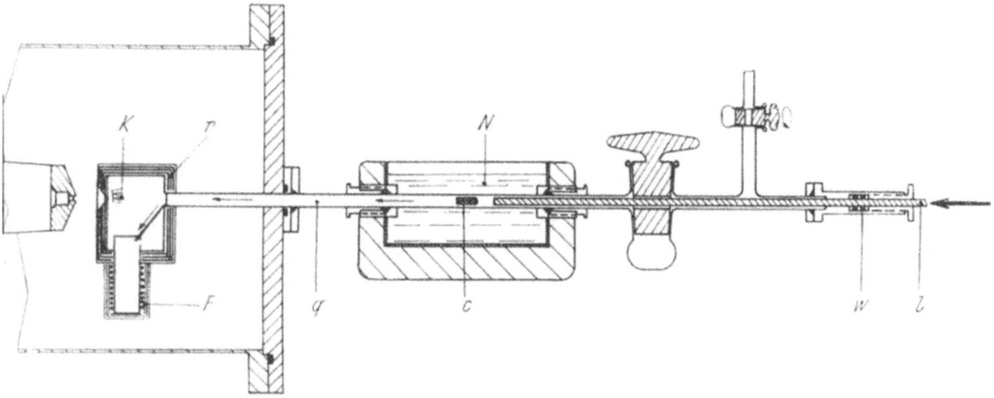

Fig. 3. Ion source for the separation of short-lived radioactive isotopes
K Filament F Oven R Heat-shields N Liquid nitrogen C Sample holder

The lock is then closed by a Wilson seal through which the rod slides and, after pumping this lock to rough vacuum, (which takes only a few seconds) it is opened and the sample holder pushed half way toward the source and stopped in a region cooled to liquid air temperature. This avoids any premature distillation of mercury.

Then the high tension is set and the mass marking done again, (approximately one minute) after which the sample is pushed into the source through an insulating rod.

After thirty seconds of collection the instrument is stopped and the thin metal foil on which the mercury has deposited can then be taken out very quickly.

Another foil can be set in place and the collection region pumped down again so that within ten minutes the instrument is ready for a new separation. The ion source can indeed receive 10 to 15 sample holders successively without needing to be cleaned out.

In order to reduce time losses, we have used at the collector a disposition which permits to count directly the radioactive isotopes deposited (Fig. 4).

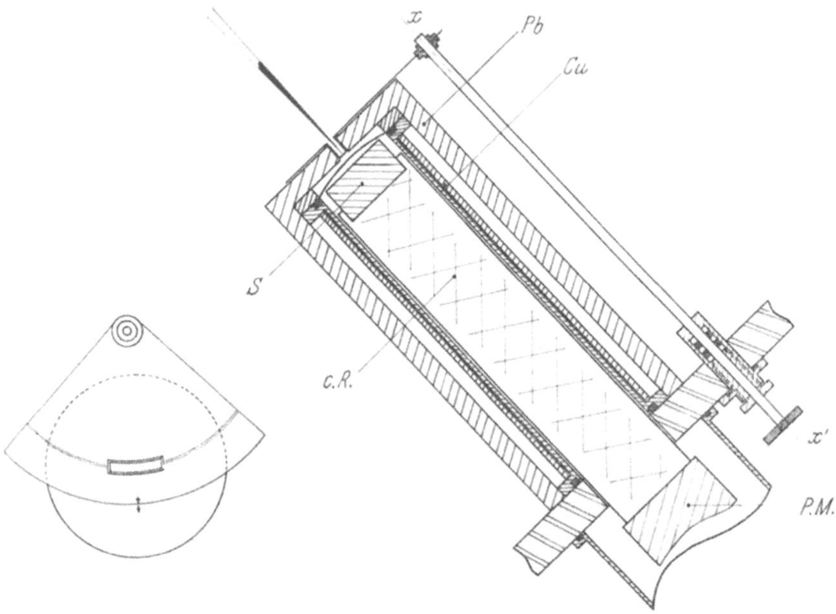

Fig. 4. Device for the measurement of the radioactive isotopes collected at the second stage

S Scintillator c. R Light guide Pb Lead shield

Results Obtained with the Source

A great number of radioactive mercury separations have been done during the last year for the study of spallation reactions.

The results have been published elsewhere [1, 2].

During these separations and with direct counting at the collector half-lives as short as about 1 minute have been studied.

Also simultaneous collections of as many as 10 radioactive isotopes have been achieved at the collector of the first stage.

In these experiments, mass marking is done with the high intensity beams for about 10 minutes on the metal foil on which the radioactive isotopes are to be deposited. Then, this foil is taken out and cut along the mass marking spots produced by the impacts so as to prepare for rapid counting of each mass after the separation.

After this, the metal foil is placed again on the collector, the tracer beams set on it again, and the radioactive sample is then sent into the source.

Extension of the Method to Elements less Volatile than Mercury

As we said before, the ion chamber temperature is about 1000° C in the experiment described. Elements such as cadmium, magnesium, and zinc are volatilised in a very short time at this temperature. This was seen very significantly when we sent into the source a small brass piece which gave rise during a short time to an intense zinc ion beam.

The field of application of the evaporation sources can then be extended considerably for such experiments.

In particular we may use for such purposes the high temperature source (which easily reaches 1600° C) and the range of elements which will evaporate very rapidly will be extended greatly.

We are in a position to say that our separator can separate without preliminary chemical treatment the radioactive isotopes of elements such as all alkalis, all alkaline-earths, magnesium, zinc, cadmium, arsenic, antimony, thallium, bismuth, and lead.

This source has also been used for the separation of gold isotopes (Au^{195} and Au^{194}) but in such a case the maximum counting rate has been obtained only after 55 minutes of collection.

Preliminary Results on a Duoplasmatron Ion Source

The duoplasmatron described by von Ardenne [3] is a hot cathode source with an axial magnetic field in which the discharge is pinched in such a way that the cathodic plasma situated in a magnetically shielded space behaves like a large cathode giving rise to an anodic plasma with high ion density.

The axial magnetic field, strongly inhomogeneous, acts on the anodic plasma in such a way as to reduce its diameter considerably in the neighbourhood of the anode where a small hole will allow the extraction of the ions.

The efficiency of such a source may reach very high values for high intensity beams. For this reason we thought it might be interesting to test it on the separator.

The source we built is shown in Fig. 5 [4].

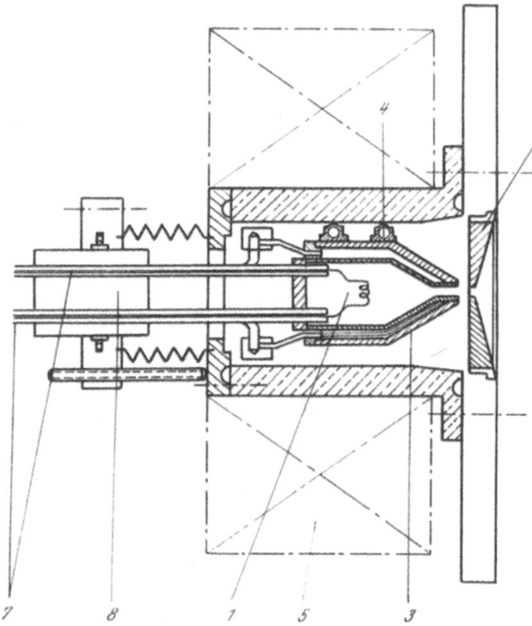

Fig. 5. "Duoplasmatron" type ion source
1 Filament 2 Anode (armco-iron) 3 Insulated ion chamber (armco-iron) 4 Quartz beads 5 Coils 7 Water cooled leads for filament 8 Insulation

First of all it was necessary to modify the ion optics of the separator to take into account the cylindrical symmetry of the ion source. A 3-element cylindrical electrode proved satisfactory.

Under these conditions losses on the wall of the machine were reduced to a minimum and true efficiencies could be measured.

Experimental results

1. *Efficiency*

In a first step this source was tested on a bench without a magnetic analyser. The measurements were made on an argon ion beam at a distance of 45 cm from the source. Efficiencies obtained reached up to 73% for a 2.5 mA beam.

These results were subject to some uncertainty due to the absence of a magnetic analyser. So we did a limited experiment in which we mounted the ion source on the separator and the results obtained, although less spectacular than the former one, are very encouraging:

collector current:	500 μA	efficiency:	8%
	190 μA		10.3%
	80 μA		44%

2. Resolving power

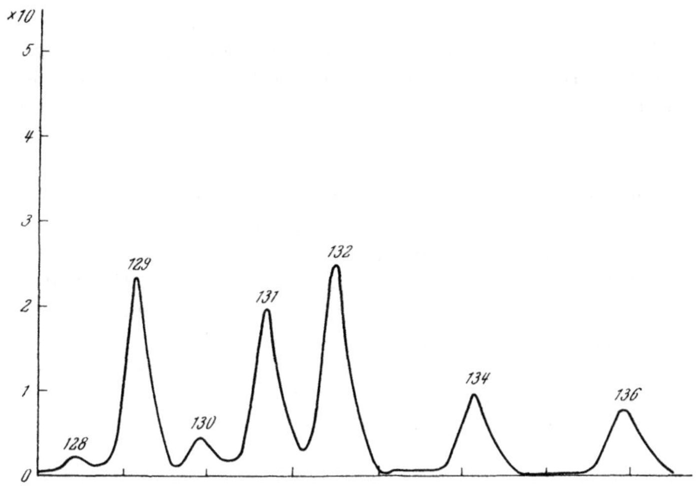

Fig. 6. Recording of the xenon spectrum obtained at the collector of the separator with the duoplasmatron ion source

Contrary to what one might have feared, the presence of a relatively high magnetic field did not destroy the resolving power. Fig. 6 shows a mass spectrum of xenon at the collector of the first stage with a resolving power close to 250.

The application of this ion source for the separation of isotopes does not seem to raise very difficult problems and if this is confirmed a few technological alterations should enable us to use it for solid materials.

A new modified source has been built and will be tried shortly.

References

1. Albouy, G., and al., C.R. Ac. Sc. **249**, 407 (1959).
2. Poffe, N., G. Albouy, R. Bernas, M. Gusakow, M. Riou, and J. Teillac, J. Phys. Radium **21**, 343 (1960).
3. Von Ardenne, M., Tabellen der Elektronenphysik, Ionenphysik und Übermikroskopie, Band I. Deutscher Verlag der Wissenschaften (Berlin 1956).
4. Klapisch, R., Thèse de 3ème Cycle (Paris 1959).

Discussion

J. H. Freeman: I would like to ask Dr. Kistemaker or Dr. Bernas what future they think there is in the duoplasmatron source for solid materials? We would like to build a duoplasmatron but the difficulty we fear is that, when the arc-chamber is heated, the iron which concentrates the magnetic field may heat above the Curie point and then the source ceases to be a duoplasmatron source. The unoplasmatron source in fact has an efficiency which is not sufficiently high to make it competitive or to make it worth while converting our present source units.

R. Bernas: We are faced with two problems. The one is to have a high efficiency source and the other is to make this source work for solid materials. If experiments on gases show that the efficiency is very high then I believe there is a future for solid materials with the same source. We built a duoplasmatron source which includes a device for high temperature production inside the source. However your remark is quite right, the iron has to be cooled to a temperature far below the Curie point. It is just a technical matter to provide a temperature gradient in the source so that the outside is cold and the inside is hot.

A Method of Rapid Electromagnetic Separation of Radioactive Isotopes

By

Jan Uhler and **Torsten Alväger**

Nobel Institute of Physics, Stockholm

With 4 Figures

In our nuclear spectroscopy work at the Nobel Institute of Physics we were faced with the problem of separating activities with half-lives as short as a few minutes. Here a short description of the technical solutions we have chosen and the main factors influencing the decisions will be given. Further, the result of some runs will be reported.

It may be pointed out that our nuclear research programme involves beta and gamma spectroscopic investigations of nuclei obtained from (p, xn) and (n,f) reactions. This means that we are generally working with samples of high specific activity, which are very often carrier-free.

From the above statements it is evident that the parameters of greatest importance are the efficiency and the separation time. For short-lived activities an increase in the efficiency and a decrease in the separation time are of course equivalent. As is well known, good efficiency is often the result of trial and error, with changes mainly of chemical compounds used and temperatures in the oven and the discharge chamber.

The normal way of separating solid activities is to mix the activities with the amount of charge material needed to adjust the instrument to stable running conditions. This means that one has always to spend time on the process of reaching thermal equilibrium in order to get good efficiency of the whole separation. This time must, of course, be short compared with the separation time. This also limits the choice of carrier to the same compound or one very similar to the active sample. All these factors set a lower limit for the separation time of about one hour.

The most satisfactory way, in our view, to overcome these difficulties is a two-furnace system and the following picture will give some information on our solution. Fig. 1 shows a picture of an ion source that has been in use for 2 years.

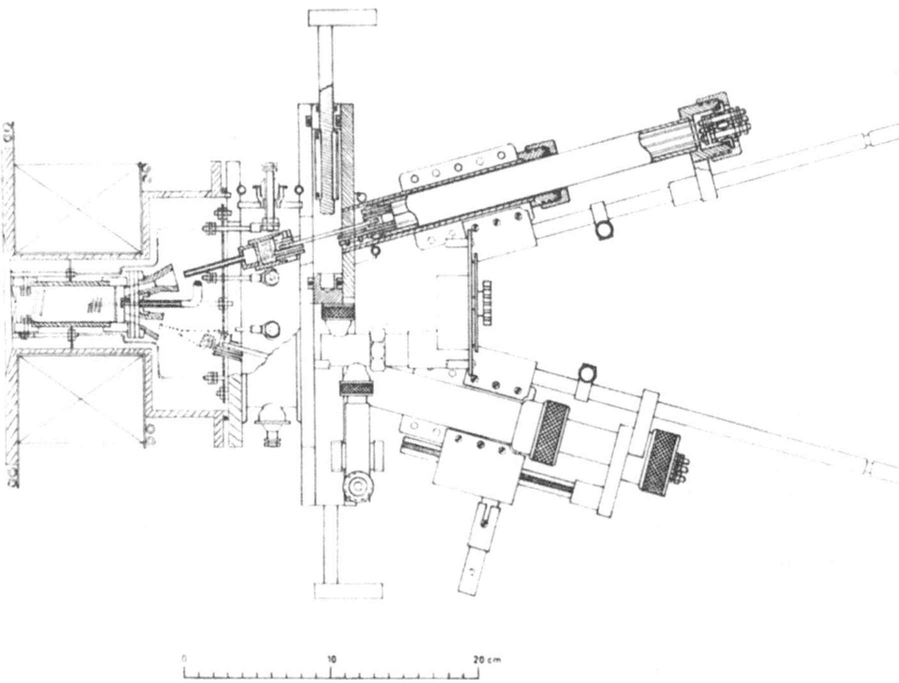

Fig. 1

The symmetrical arrangement makes it possible to supply the ion source with two identical furnaces, which may be suitable for some separations. The two furnaces are both indirectly heated. The stable-material furnace is made of stainless steel with an inner container of graphite. The active-material furnace is of platinum, chosen for mechanical reasons. Both furnaces are fastened by long stainless steel rods to brass tubes. Conductors go through these tubes for the heating current and thermocouple. Because of the small volume of the tubes and a good backing pump, the furnaces can be put into the main vacuum in less than one minute without destroying the high vacuum necessary for operation.

The discharge chamber, shown in Fig. 2, is built of separate, easily machined parts of stainless steel, molybdenum, pyrophylite, and aluminium oxide. These materials give low gassing after air has been in the ion source. At higher temperatures chlorine combines with the steel. Dimensions and some details are about the same as used by the other Scandinavian groups. The chamber 1 is a cylinder of in-

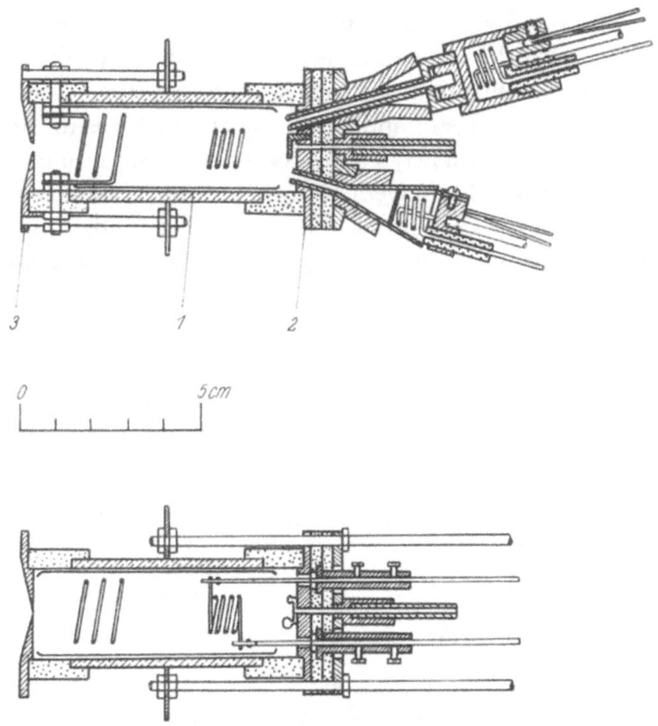

Fig. 2

sulating material, with end plates, 2 and 3, of stainless steel. Cover plates of graphite are often used for these. Inside the insulating cylinder there is a molybdenum tube which is the anode. This arrangement makes it possible to use a discharge of either a low voltage arc or an oscillating electron type. The cathode, a tungsten spiral is suspended between molybdenum rods which are fastened to holders of stainless steel. As it has been found that a metallic deposit is formed by the cathode, the insulating parts of the cathode holders are in shadow with respect to the tungsten wire. A separate heater is needed in order to obtain higher temperatures independent of the cathode effect. The anode and the cathode are fed by direct current. At the opening for the active material furnace there is a small molybdenum shutter.

A separation is performed as follows: After preheating and starting the separation on stable material it takes at least half an hour to reach operating conditions. During this time any chemical procedures used are finished. The solution of the active sample is then put into the platinum furnace and evaporated to dryness. Then the high voltage supply to the ion source is turned off, with the discharge still running, and the platinum furnace is put into the main vacuum. Separation is started again and, after a check of focusing properties, the activity furnace is slowly connected to the discharge chamber by means of a mechanical arrangement with a motor. The cold platinum tip gives a decrease in pressure for about 10 seconds, but then the pressure rises again and the heating of the activity oven can start.

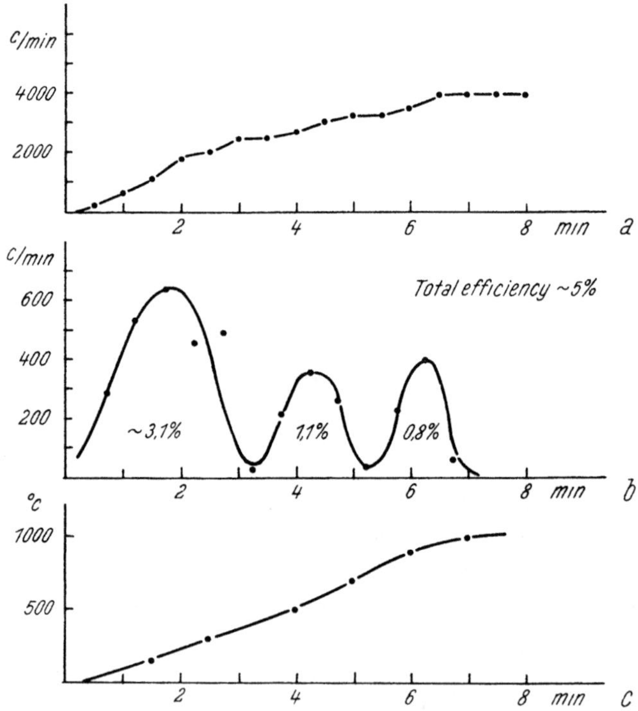

Fig. 3

As a test, carrier free Pb^{212} has been separated as $PbCl_2$. To investigate the change in activity transfer during the time of the separation, a beta-sensitive counter is mounted in the target chamber.

The counting rate measured at half minute intervals is shown in Fig. 3a.

The difference in the counting rate between successive intervals is given in Fig. 3b. To get an approximate measure of the temperature in the platinum furnace, a thermocouple is mounted in the stainless steel plate supporting the platinum part. The temperature rise is given in Fig. 3c.

A very rapid separation has been performed by utilizing the possibility of introducing an oven with low heat capacity during separation. Iodine was collected on a silver foil by simply dipping the foil into an iodine solution. With xenon as charge material, the furnace containing the silver foil was connected to the discharge chamber and the effective separation lasted for about one minute.

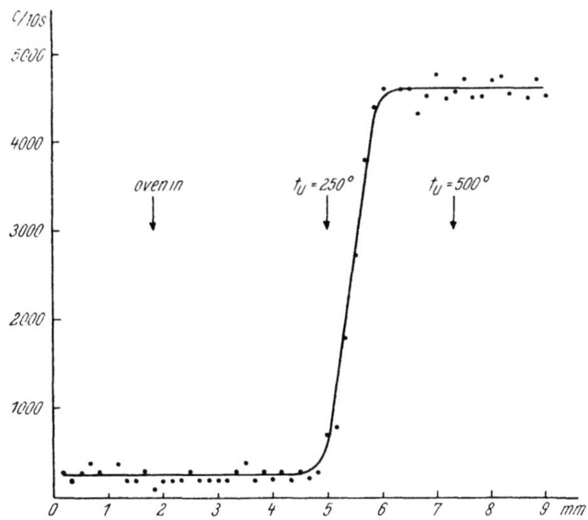

Fig. 4

The efficiency here was about 3%.

Reference

1. Uhler, J., and T. Alväger, Ark. Fys. **14**, 473 (1958).
2. Alväger, Thesis (1960).

Discussion

R. Bernas: Dr. Uhler mentioned that for the Iodine separation, the actual separation time was of the order of 1 minute. What was the total time between the moment the source was put into the machine and the moment the target was pulled out?

J. Uhler: The total time between receiving the sample from the chemical laboratory and taking the target out of the machine was about 4 minutes.

O. B. Nielsen: I would like to point out that the important time when evaluating the different methods is the time passing from the moment the activity comes from the cyclotron until the separation is finished and the time needed for the production of the carrier free activity.

J. Uhler: At the separation of alpha active elements produced by heavy ion bombardments in the cyclotron some of the chemical procedures were effected in the elevator from the cyclotron to the separator. In that way it was possible to measure half-lives and energies of alpha-particles with a half-life of 5 minutes.

Some Problems Related with Ion Sources for Radioactive Isotopes

By

P. Lazeyras
C.E.N. Saclay, France

With 3 Figures

I. Introduction

When processing stable elements in an E.M. Separator, the chief requirements for the source operation are steadiness, high resolution, high currents, in certain cases, and a long life of the cathode. Efficiency is very often a secondary requirement, at least for low scale production, even when the element or the compound is expensive. In the same way, the rapidity of the source operation is seldom important.

When processing radioelements of a short-life, the efficiency and the rapidity of the operation become chief requirements. On the other hand the life of the cathode and the high currents can be sacrificed. It appears then that the design and the operation of the sources for short-lived radioisotopes can differ from those for stable isotopes or long-lived radioisotopes.

The efficiency of an ion source is the ratio between the number of atoms of the processed element leaving the source as unitary charge atomic ions and accelerated as a homogeneous beam, and the number of atoms of the same element present in the source before the source is run.

Setting aside the atoms which are lost when initiating the run of the source, we can define the internal efficiency which corresponds to atoms lost inside the source: e.g. dark points the cold points, and the external efficiency, the latter being the overall source efficiency when the processed element is volatile. It corresponds to atoms lost outside the source; neutrals crossing the exit slit for ions, side bands, etc.

The rapidity of the seperation must be great but can be adjusted in connexion with other operations (principally chemical ones). All these requirements can vary according to the period of the processed nuclide.

a

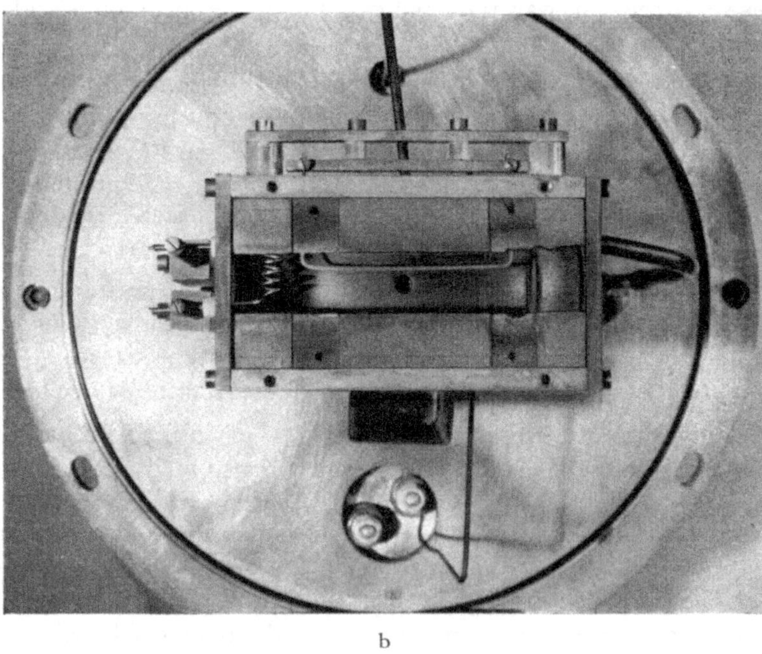

b

Fig. 1. Ion source for radioisotopes
a General view b Front view, the front plate having been taken off

II. Description of an Ion Source for Short-Lived Radioelements

The elements we intended to process being in the rare earths group, we feared the internal efficiency would be exceedingly small if convenient precautions were not taken (Fig. 1).

This source works on the sputtering arc principle. The arc being run with the help of a support gas (argon in almost all the cases) a negative electrode is set inside the discharge and is sputtered by the support gas ions. The sputtered atoms are in turn ionised in the discharge. In order to avoid the redeposit of these atoms onto dark points, that is, corners or an anodic surface, the arc chamber construction avoids the corners as far as possible and its potential is held at cathode potential. The anode is a tungsten wire which runs very hot (tantalum anodes melt easily) and on which no rare earth atom condenses, as was directly verified. The sputtering voltage is the main arc voltage, which can be made larger than usual ones (e. g. 300 volts instead of 60 volts), in order to raise the sputtering ratios; the drawback is a shorter life for the filament cathode. But this drawback is of no importance since the separations do not last longer than 30 minutes at maximum.

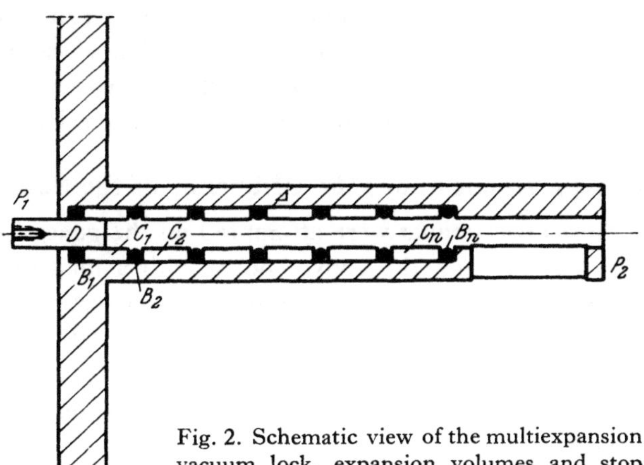

Fig. 2. Schematic view of the multiexpansion vacuum lock, expansion volumes and stop

In order to introduce the sample from the main laboratory into the source, a vacuum lock was designed for the following requirements: no auxiliary pumping, no disturbance in the source and the whole separator regime, and rapidity in operation.

The sample is located in a small case drilled into the metal rod. The operation consists in introducing the sample without stopping the run while the separator is running with guide beams. There is a multi-expansion lock in which the air is stacked (it is released inside the separator after the separation is over). The stopper is held in place by the tightening of the rubber gaskets and drops into a box through a convenient aperture. The sample holder goes on its way since it is screwed onto the driving rod. When the latter is pulled back, after the separation is over, the air is released and the sample holder is unscrewed when partly emerging into the open air. Thus it can be used as a stopper for a subsequent separation.

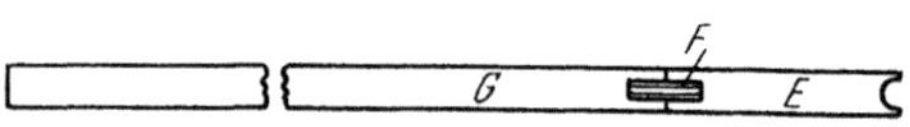

Fig. 3. Sample holder and driving rod

This source was run in a few separations of rare earth material irradiated in a cyclotron (neodymium and praseodymium). Separation of an isotope of prometheum (mass 141, period about 22 minutes) was intended.

The basic principles of the design of the above source were found to be satisfactory; the internal efficiency appeared to be high. There was very little radioactivity inside the source after a run and the lock worked perfectly. Nevertheless the overall efficiency was very poor, which means that we must try to make further improvements, though the possibility of faulty operation should not to be overlooked.

III. Experiment about the Efficiency of the Sources

A few short experiments were attempted on the general efficiency problem. When running the above source with a mixture of argon and xenon gases, the flow of each compound being separately adjustable and separately measured, the efficiency can easily be measured for each component by the help of current measurements. These efficiencies were found to be roughly steady; when the relative flows of gases varied from the ratio 1/10 to 10/1 then each component seemed to behave in the source as if the other one were not present. This is a recent work and no comment can yet be given.

IV. Efficiency and Resolution

Resolution must on no account be sacrificed. It then appears that the actual values of efficiency and resolution are matters of compromise.

As a matter of fact, I state that when testing conventional ion sources on low resolution experimental setups, efficiency can be as high as 30%. There are several similar results in literature and I observed it myself. When transferred into the main separator, the corresponding regime of the source leads to hashy beams without any resolution. The actual efficiencies of our separator with stable elements are about a few per cent. So it appears that the more hashy the arc, the higher the efficiency and a suggestion will be made now, although no experimental attempt has yet been made for the verification. In the lecture on the isotopic contamination it was stated that the contamination due to the hash could be represented by the formula $c = \frac{\mu i}{p}$. When the efficiency and the hash are high, the coefficient μ is also great, but the whole term can be reduced to reasonable values by lowering the intensity. It would be possible to keep good resolution and good efficiency by reducing the width of the source slit, that is at the expense of lower intensity and longer duration of the separations. It may happen that in some cases this compromise is favourable.

Another problem which can be set is as follows. In Saclay we try to process radioactive elements with our available separator. To the source arc we can apply only auxiliary fields of about 100 Gauss. For stable nuclides, the efficiency is of the order of 1 to 2%. It is seen in the literature that in calutron operation efficiency can be much greater, which is due partly to better profiles in the dissociation spectra of compounds (the side bands are weaker in a calutron operation) and partly to poorly known gas discharge phenomena.

It would be of interest to find out by subsequent research whether an auxiliary magnetic field of low intensity would be convenient for the compromise resolution-efficiency, or whether the sector principle must be abandoned in new instruments for radioisotope studies, where this requirement is of prime importance. It is seen from private communications that the flexibility of arc sources is not good, namely below 2000 G for the applied magnetic field. Further information would be necessary.

A Parallel Filament Lateral Arc Source Unit

By

J. H. Freeman

Chemistry Division A.E.R.E., Harwell

With 2 Figures

In an attempt to overcome the difficulties associated with the Hermes source unit [1] an electron impact source unit with the filament behind and parallel to the extraction slit and parallel to the focusing magnetic field, has been tested. It was hoped that in this arrangement

a) the arc could only take up one position in the arc chamber,

b) an intense arc close to the extraction slit would be produced,

c) the magnetic field due to the filament and the source magnet field would produce a resultant spiral field around the filament and so give a very long electron path.

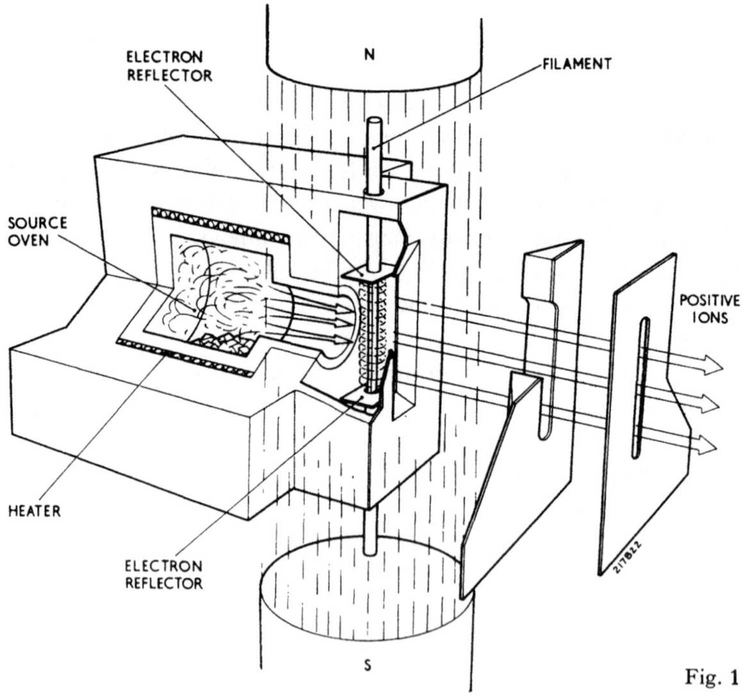

Fig. 1

The layout of the source is shown diagramatically in Fig. 1. The electron reflectors were introduced to increase the electron path still further.

Preliminary experiments made on a crude experimental assembly just before the conference were very promising. With a strip tantalum filament (4 × 1 × 50 mm) placed 3 mm behind the extraction slit, large well focused beams of krypton and argon were obtained. Quite contrary to normal experience the focus remained excellent over almost the whole range of outputs and arc conditions. An excellent resolution was also obtained with the source magnet switched off altogether. Comparative measurements indicated that the efficiency was approximately twice that of the conventional source unit. After 40 hours of operation under a wide variety of arc conditions the filament showed little signs of wear. The source was also tested with *very* thin strip tantalum foil and 1 mm tantalum wire cathodes and gave results similar to those described above.

Reference

1. McIlroy, R. W., J. H. Freeman, and K. J. Hill, p. 155. Present volume Recent Modifications to Hermes.

Remark

by W. Walcher, Marburg

Dr. Freeman has reported on the distortion of the magnetic field in the ion source by the field of the cathode filament. We have constructed

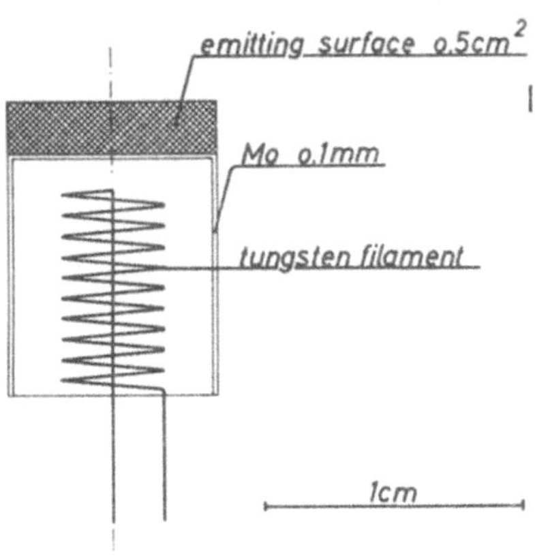

Fig. 2

an indirectly heated cathode type which does not show such disturbance. Fig. 2 shows the arrangement and Table 1 the operating data.

Table 1

Th + 3% Ba				Th + Mo (3 : 7)
U_{Arc} = 40 V	20 V	100 V		U_{Arc} = 100 V
t_{life} = 24 h	48 h	10 m		t_{life} = 45 h
N_H = 30 W				N_H = 150 W
T = 1200° K				T = 1700° K
j_{Arc} = 1 A/cm²				j_{Arc} = 1 A/cm²
N_{Arc} = 20 W				N_{Arc} = 50 W
$N_{Arc} + N_H$ = 50 W				$N_{Arc} + N_H$ = 200 W

Chapter V

Sputtering Phenomena

Sputtering Experiments with the Amsterdam Electromagnetic Isotope Separator

By

P. K. Rol, J. M. Fluit, and **J. Kistemaker**

F. O. M.-Laboratorium voor Massaspectrografie, Amsterdam

With 10 Figures

Summary

The sputtering ratio of polycrystalline copper is measured by bombarding with ions of many elements as a function of the ion energy and the angle of incidence. For copper monocrystals the sputtering ratio has been measured for different crystal orientations as a function of the angle of incidence in bombardment with Ar^{+}-ions at 20 keV. An interpretation of the measurements is given on the basis of rigid sphere collisions of the ions in the lattice. It is supposed that excited regions, which emit the sputtered particles, are formed near the surface.

1. Introduction

In the collection of separated isotopes in electromagnetic isotope separators many problems arise and many strange effects are not yet understood. Besides the pure scientific interest this is one of the reasons that sputtering research with the Amsterdam separator was started [1, 2]. In much experimental work the sputtering process has been studied by bombarding different target materials with ions of only a few elements. By changing the target materials, however, many more parameters are changed than by using ions of different elements and one target material. An interpretation of the results in the first case is therefore very difficult. For this reason only copper was used as target material in our experiments. For these sputtering experiments a copper target was mounted at the place where normally the receiver for the separated isotopes is located. As the target is placed in the strong magnetic field no correction should need to be made on the ion current for secondary emission of electrons, but, due to edge effects and the small negative potential of the target, a small correction on the ion currents must be made.

The targets with dimensions of $90 \times 20 \times 1$ mm were made of electrolytic copper. On the back of the target was soldered a strip by which it was clamped on a water cooled tube. The monocrystalline targets were pressed by a graphite diaphragm on a copper backing plate which was again clamped on the tube. The maximum temperature of the target was always below 125° C. In order to measure the sputtering ratio for different angles of incidence, the target can be clamped in different positions on the tube in such a way that the target is always parallel with the magnetic lines of force to suppress the secondary emission. The target is surrounded by a diaphragm, which makes it possible to measure only the current on the target and not on the tube and the graphite diaphragm.

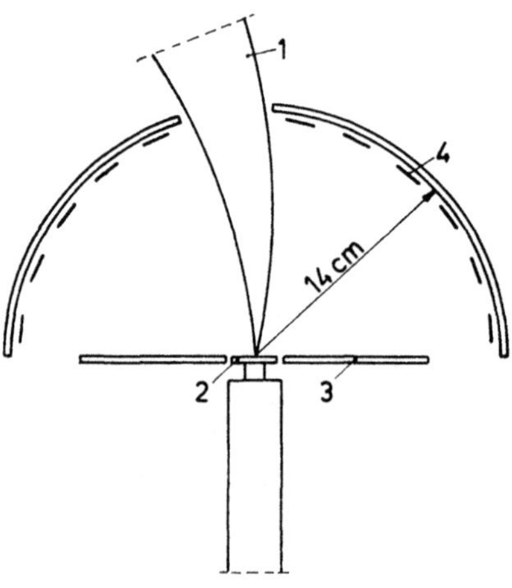

Fig. 1. Schematic diagram of the experimental set up
1. Beam, 2. Target, 3. Diaphragm, 4. Glassplate

The polycrystalline targets were mechanically polished and cleaned thoroughly with trichloroethylene, alcohol, and ether. The monocrystalline targets were only electropolished. The sputtering ratio, that is the number of atoms released per impinging ion, can be calculated from the total charge of the ions reaching the target and the loss in weight.

The detection of the sputtered material in different directions was done by condensation of the copper atoms on small glass plates surrounding the target. From the increase in weight of these plates it was possible to calculate the relative intensity of copper atoms emitted in the different directions.

2. Experimental Results for Polycrystalline Copper

a) The sputtering ratio

The sputtering ratio was measured for copper bombarded with single charged ions of He, Li, C, N, O, Ne, Na, Mg, Al, Si, P, S,

Cl, Ar, K, Ca, Cu, Zn, Sr, Zr, Cd, J, Hg, and Tl. Not in all the cases, however, could we get reliable values for the sputtering ratio. In some cases a layer was built up on the target which could resist sputtering. This was especially the case for ions of those elements which form very stable oxides, namely Mg, Al, Ca, Sr, and Zr, while oxygen ions also formed a blue layer on the surface. When copper was bombarded with ions of these elements, the sputtering ratio was always very low and sometimes the target even increased in weight. The sputtering ratio was also very low in the case of bombardment with C^+-ions, but here a very hard layer of a carbon-copper mixture was found on the surface. Very low sputtering ratios for He^+- and Li^+-bombardment may also partially be due to surface layers, the sputtering being so low, that it cannot compete with the formation of oxydes.

In Si^+-, P^+-, and S^+-bombardment, part of the impinging ions were trapped in the target, while with the other elements no traces of the impinging material could be detected in the target. All the values of S have to be corrected properly, for besides the copper the impinging material is also sputtered. To get a reliable estimate for this correction, one has to know the relation between the crystal parameters and the sputtering. Not enough information is available, however, so this correction is not made except for copper.

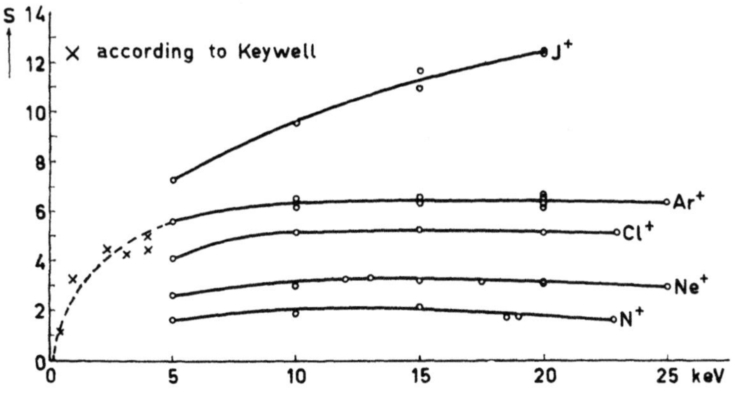

Fig. 2. Sputtering ratio of copper by bombardment with N^+-, Ne^+-, Cl^+-, Ar^+-, and J^+-ions for normal incidence

The results for the sputtering ratio of the other elements as a function of energy are given in Figs. 2, 3, and 4.

The divergent behaviour of Hg^+-ions may also be due to surface contamination of the copper by mercury vapour.

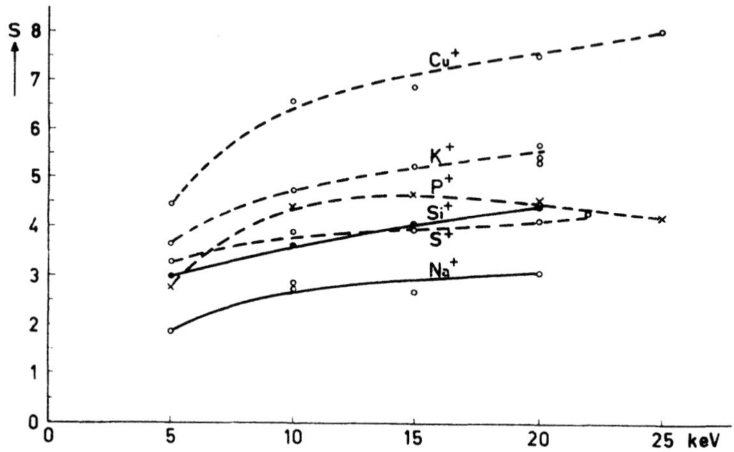

Fig. 3. Sputtering ratio of copper by bombardment with Na^+-, Si^+-, P^+-, S^+-, K^+-, and Cu^+-ions, for normal incidence

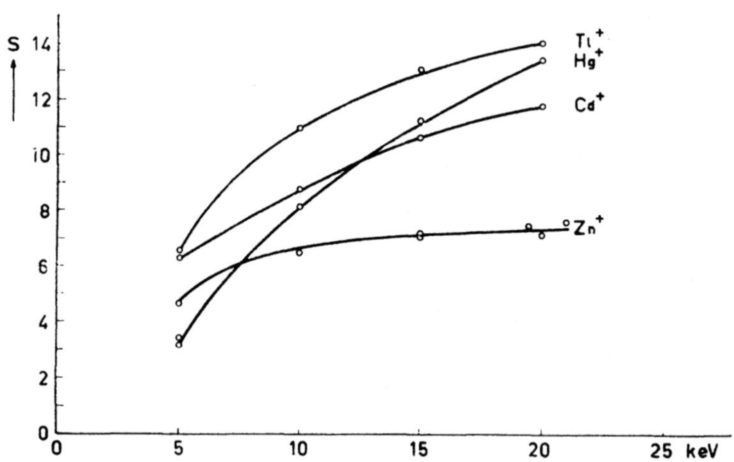

Fig. 4. Sputtering ratio of copper by bombardment with Zn^+-, Cd^+-, Hg^+-, and Tl^+-ions, for normal incidence

b) The angle of incidence

When copper was bombarded with Ar^+- and Tl^+-ions the sputtering ratio was measured as a function of the angle of incidence at 20 keV.

Very high sputtering ratios were measured here for Tl^{+}-ions at oblique incidence. The curves in Fig. 5 are characteristic, however, for one energy only. At lower energies the increase is somewhat less.

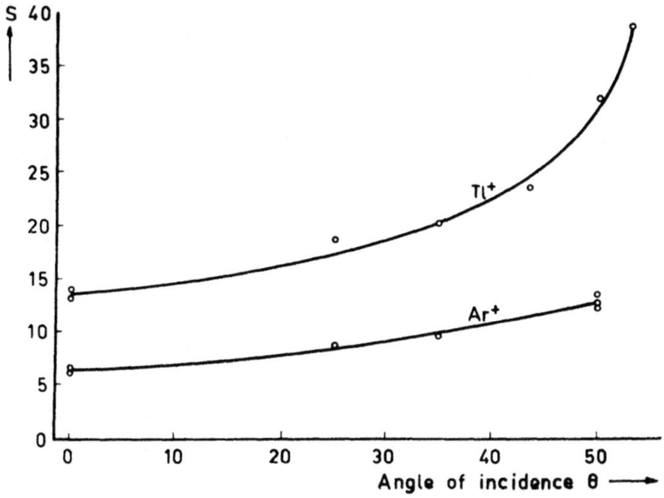

Fig. 5. Sputtering ratio of copper as a function of the angle of incidence by bombardment with Tl^{+}- and Ar^{+}-ions at 20 keV

c) Angular distribution of the sputtered material

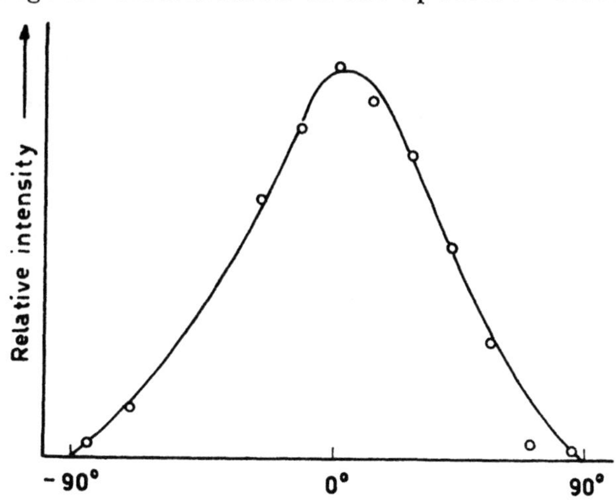

Fig. 6. The angular distribution of sputtered copper by bombardment with Ar^{+}-ions of 20 keV. Angle of incidence was 50°

The angular distribution measured by SEELIGER and SOMMERMEYER [3] could be described by a cosine function. The two important results found by us are:

1. the distribution was a Gaussian rather than a cosine distribution;
2. the distribution is always, even for oblique incidence, symmetrical with respect to the normal on the target surface.

3. Experimental Results for Monocrystalline Copper

a) The sputtering ratio

The sputtering ratio of copper monocrystals was measured by bombarding with Ar^+-ions at 20 keV. The sputtering ratio at normal incidence was 8.2 for a crystal where the surface coincided with a (111) crystal plane, 3.2 for a (100) surface and 2.8 for a (110) surface, while we measured 6.3 for polycrystalline copper under the same conditions. From these results we see that even at these high energies of the ions, the monocrystalline nature of the surface exists after a long time of bombardment. During this time layers of 10 μ have been removed by sputtering.

b) The angle of incidence

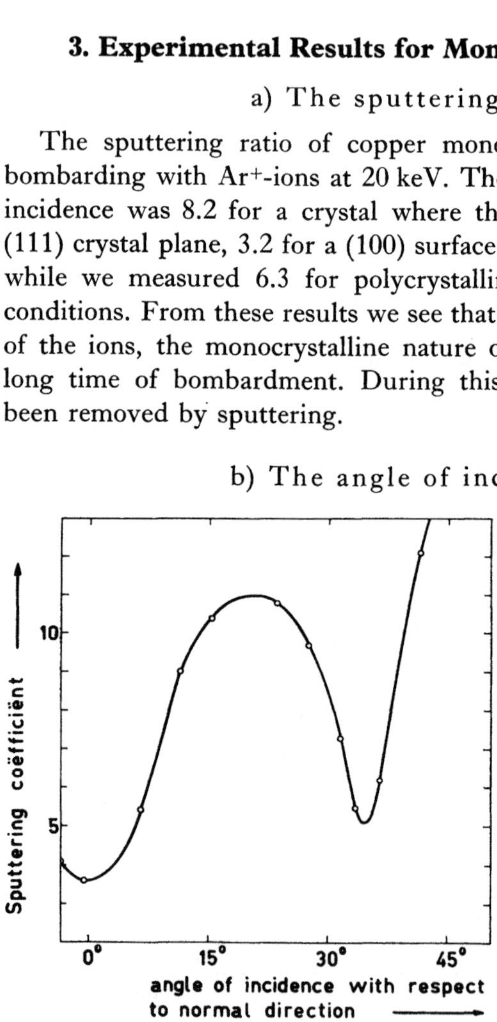

Fig. 7. The sputtering ratio for a (100) surface of a copper crystal, which is turned around a [011] direction in the surface

Also here the difference in the results for polycrystalline copper is considerable. It is important around which axis the crystal is turned, as can be seen from Figs. 7 and 8. In both cases the sputtering ratio was measured with Ar^+-ions of 20 keV. In Fig. 7 the crystal is turned around a [011] direction in the surface and in Fig. 8 around a [001] direction. These figures have to be compared with Fig. 5 for polycrystalline copper. In the curve of Fig. 7 there are perhaps not enough measuring points to give a fine-structure such as in Fig. 8.

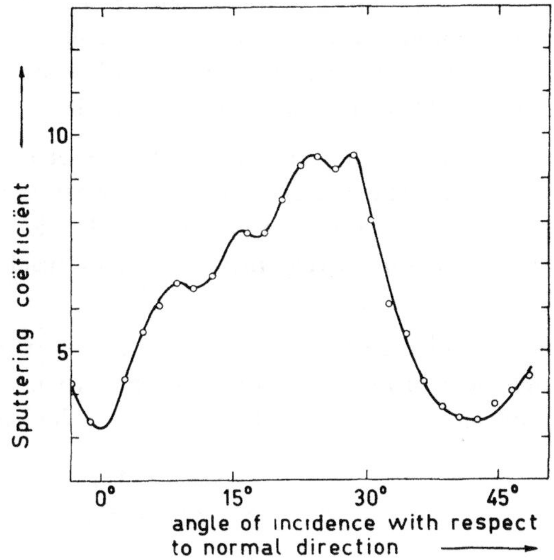

Fig. 8. The sputtering ratio for a (100) surface of a copper crystal, which is turned around a [001] direction in the surface

c) Angular distribution of the sputtered material

The angular distribution was measured here with a hollow glass balloon. The beam reached the target through holes in the balloon. The target was mounted at the exit hole of the beam and the bombarded surface coincided with the inner target surface of the balloon. The relative thickness of the copper layer on the sphere was investigated by measuring its transparency with a photocell. The thickness on different places allowed calculation of the relative intensity of the sputtered particles in the different directions.

When bombarding the (111) crystal, we found spots on the glass balloon in the three [110] directions, which did not lie in the surface, like Wehner [4]. However, spots in the three [100] directions were also found. These had about half the intensity of the spots in the [110] directions. When bombarding the (100) crystal we found spots in four [110] directions. Yurasova [5] has measured that the angular distribution does not change for oblique incidence.

4. Discussion

The agreement of the experimental results with any of the existing theories is poor. The theory of Bohr [6] on penetration of energetic particles into matter was used, in order to get an insight into the basic processes which occur in the sputtering process [2].

The rigid sphere model can be used in the energy region under consideration. For collisions with small angle scattering this model may not be used but, as these collisions give rise to small energy transfer, they are not important for the sputtering process.

It is likely that only the collisions of the ions in the lattice which take place near the surface are responsible for sputtering, that means only the first collisions of the ions. Particles released deep under the surface will have very little chance of reaching the surface and being sputtered.

The chance that an ion will have a collision near the surface is inversely proportional to λ, the mean free path for elastic collisions. The energy transferred in this first collision is proportional to

$$\frac{M_1 M_2}{(M_1+M_2)^2} E$$

where M_1 and M_2 are the mass numbers of the ion and target atom respectively and E is the energy of the ion. The angular distribution of the particles released from their lattice sites by the ion is independent of M_1 and M_2.

In first order approximation we can suppose

$$S \sim \frac{1}{\lambda(E)} \frac{M_1 M_2}{(M_1+M_2)^2} E$$

For oblique incidence the apparent mean free path of the ion in the first atomic layers is $\lambda(E) \cos\varphi$, where φ is the angle of incidence. This gives

$$S = K \frac{1}{\lambda(E) \cos\varphi} \frac{M_1 M_2}{(M_1+M_2)^2} E$$

where K is a constant for one target material.

Here it is supposed that a region of the target near the surface is excited, and from this excited region the atoms are emitted from the metal, while the number is proportional to the amount of energy transferred. The collision radius for the rigid sphere model can be calculated according to BOHR [6] for a screened potential as

$$R = C \frac{a_0}{(Z_1^{2/3} + Z_2^{2/3})^{1/2}} \ln \frac{Z_1 Z_2 e}{\varepsilon_0 R E'}$$

where $E' = \frac{M_2}{M_1 + M_2} E$, C is a constant, a_0 is the "radius of the hydrogen atom", Z_1 and Z_2 are the atomic numbers of the ion and the target atom respectively, e the elementary charge and ε_0 the dielectric constant in vacuum. The mean free path is then given by

$$\lambda = \frac{1}{\pi R^2 n_0}$$

where n_0 is the number of lattice atoms per unit-volume. The calculated values for the sputtering ratio of copper by bombardment with ions of several elements are given in Fig. 9, where K was chosen as

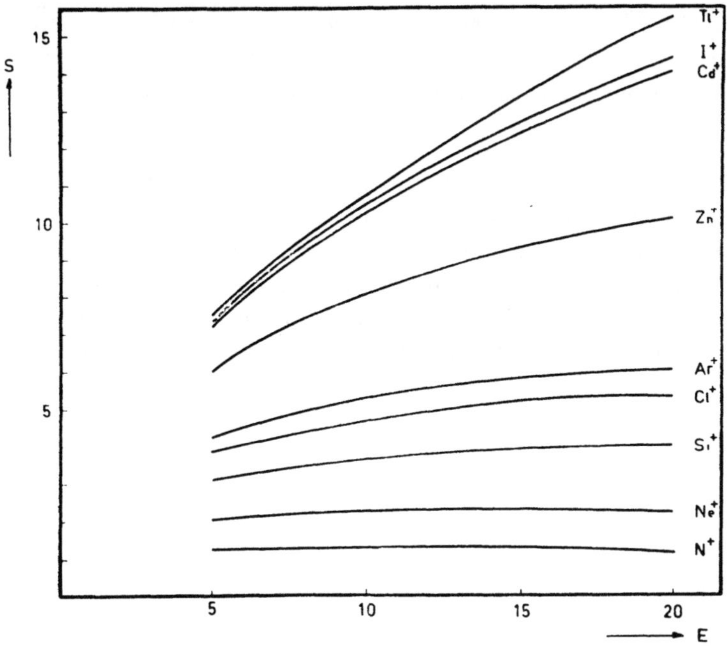

Fig. 9. Calculated values for the sputtering ratio with $K = 1.67 \cdot 10^{-11}$ m/eV

$1.67 \cdot 10^{-11}$ meter per eV giving the best agreement with the experimental results and C was 1. The tendency of these curves is in accordance with the experimental data, especially for every element individually as can be seen from Fig. 10, where we plotted the ratio of the calculated sputtering ratio to the measured one as a function of the ion mass.

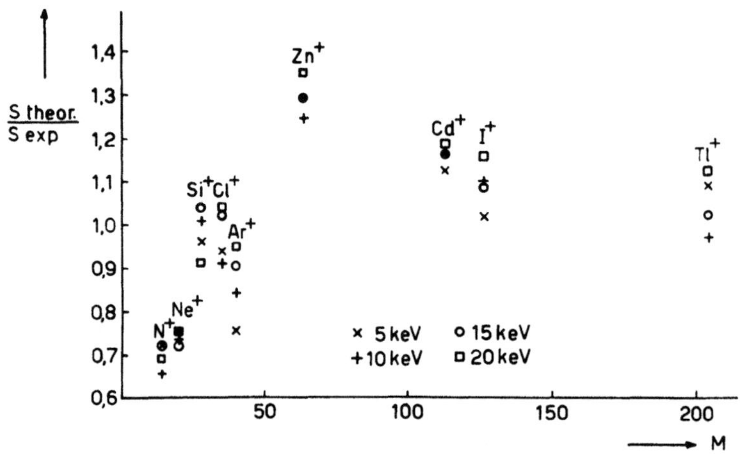

Fig. 10. The ratio of the calculated and measured sputtering ratios as a function of the ion mass

The discrepancy between theory and experiment can partly be understood from the approximations made in the theoretical model. The high experimental values for the lighter elements may be due to the fact that by back-scattering of the ions more collisions take place near the surface and the small angle scattering is more important for these elements than is taken into account. The higher experimental values for the heavier ions can be understood by assuming that more collisions take place near the surface, as the mean free path is here rather small.

This model of sputtering can also explain the results of Fig. 7 and Fig. 8. Minima for the sputtering ratio are found for those directions in which the crystal is transparent, that means the mean free path is great. The maxima are found for directions where the crystal is nontransparent so the mean free path of the ion is small [8]. The crystal is also rather nontransparent in the [111] direction; the transparency is high in the [100] and [110] directions and highest for the last one. This is in accordance with the measured values for the sputtering ratio. Calculations for the sputtering ratio as a function of the transparency have not yet been made. The sputtering ratio for polycrystalline copper increases more with the angle of incidence than is given by the formula. Furthermore, the theoretical model does not predict the angular distribution of the sputtered material as the mechanism of excitation is not further investigated. Other experiments with monocrystals have to be done to analyse this mechanism.

5. Acknowledgments

The authors wish to thank Prof. Dr. G. W. RATHENAU for his stimulating interest. They are indebted to N.V. PHILIPS, Eindhoven, Holland, for placing the monocrystals at their disposal and to Mr. M. DE JONG for preparing the monocrystals.

This work is part of the research program of the Stichting voor Fundamenteel Onderzoek der Materie and was made possible by financial support from the Nederlandse Organisatie voor Zuiver Wetenschappelijk Onderzoek.

References

1. ROL, P. K., J. M. FLUIT, and J. KISTEMAKER, Terzo Congr. Int. sui Fenomeni d'Ionizzazione nei Gas. Soc. It. di Fis., p. 872 (1957).
2. ROL, P. K., Thesis, Amsterdam 1960.
3. SEELIGER, R., and K. SOMMERMEYER, Z. Physik **93**, 692 (1935).
4. WEHNER, G. K., J. Appl. Phys. **26**, 1056 (1955).
5. YURASOVA, V. E., N. V. PLESHIVTSEV, and I. V. ORFANOV, J.E.T.P. **37**, 689 (1960).
6. BOHR, N., Kgl. Danske Videnskab. Selskab. Biol. Medd. 18, No. 8 (1948).
7. ROL, P. K., J. M. FLUIT, F. P. VIEHBÖCK, and M. DE JONG, Ionization Phenomena in Gases, Edited by N. R. NILSSON, North-Holland Publ. Co., p. 257 (Amsterdam 1960).

Discussion

J. H. FREEMAN: Dr. ROL mentioned in his paper that the sputtering distribution appeared to be Gaussian rather than a cosine distribution. It might be fundamental for sputtering theories to know the exact distribution because there is a possibility of an evaporation process involved.

P. K. ROL: The Gaussian distribution is not very well marked. There are measurements from O'BRIAIN and COBIC [1, 2] who found in the very beginning of the sputtering process a cosine distribution; when the target is more and more roughened the distribution becomes more peaked in the forward direction but it is always symmetrical around the normal to the target surface. But I don't think that one can conclude from these facts that sputtering is an evaporation or an impulse effect. At low energies

WEHNER [3] found that at oblique incidence the particles come off more in the reverse direction from the impinging beam but at high energies we could not find such an asymmetry.

J. H. FREEMAN: If there is a cosine distribution would it not be changed by the fact that the height of the sputtering slit is comparable with the distance from the sputtering source to the collector.

P. K. ROL: No, this can be calculated. If there is originally a cosine distribution we should have measured the cosine distribution.

K. O. NIELSEN: I am surprised that the sputtering coefficient "*S*" is not dependent on the mean free path of the target atom in the target material.

P. K. ROL: The mean free path of the target atoms does not come in the equation because it is the same for all bombarding elements. The physical picture is that an excited region is developed by the impinging ions near the surface and from this excited region particles are emitted and sputtered. The mean free path will always be the same and, therefore, we made all these experiments with copper.

K. O. NIELSEN: The expression "rigid sphere model" should not be used for collision in this energy region. In the lower energy region one has the real "rigid sphere model," that means collision where the cross-section is independent of the energy. In the energy region we have here, we should use the term "isotropic scattering."

P. K. ROL: For our treatment we introduced the "rigid sphere model" assuming that the radius of the "rigid sphere model" was energy dependent. This "rigid sphere model" was also used by SEITZ in his theory on radiation damage.

References

1. O'BRIAIN, C. D., c.s. J. Chem. Phys. 29. 3. 1958.
2. COBÍC, B., c.s. Ionisation Phenomena in Gases, Ed. by N. R. NILSSON, North Holland Publ. Co., p. 260 (Amsterdam 1960).
3. WEHNER, G. K., Adv. in Electr., Ed. by L. MARTON, VoL. VII. Acad. Press, p. 239 (New York 1955).

Collection and Sputtering Experiments with Noble Gas Ions

By

O. Almén and **G. Bruce**

Department of Physics, Chalmers University of Technology, Gothenburg

With 9 Figures

The collection and sputtering experiments described in this paper were made with the 90° electromagnetic isotope separator in our laboratory [1].

1. Experimental Arrangement

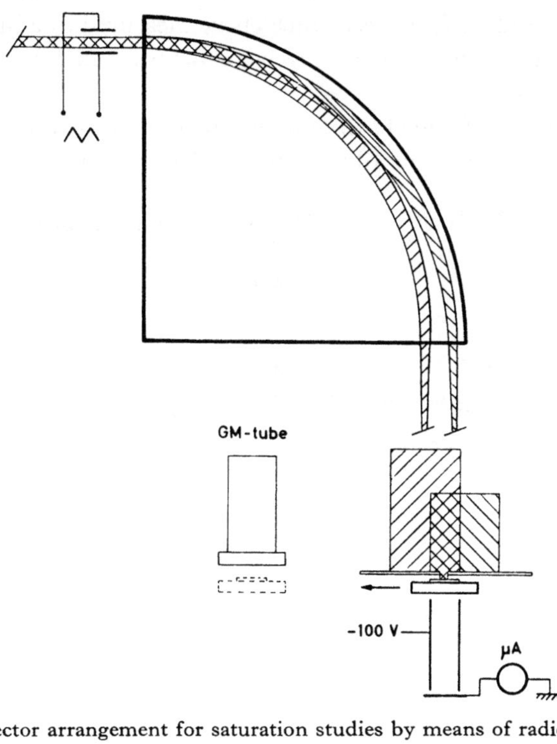

Fig. 1. Collector arrangement for saturation studies by means of radioactive tracers

To study collection of Kr ions we used an arrangement, schematically shown in Fig. 1. Natural Kr, mixed with the beta-active

Kr^{85} (half-life 10 years), is swept over the target by means of a saw-tooth voltage on the deflecting plates. Where the two beams Kr^{84} and Kr^{85} overlap, collection is studied on a plate situated behind a circular aperture. The target plate can be moved to a position in front of a G.M. tube, where the activity is determined. At the same time the ion current is measured in a Faraday cage. After saturation is reached the activity is compared outside the vacuum chamber with that of a reference target containing a known collected quantity. In this way we can measure relative Kr amounts with an accuracy of 2–5 per cent.

Determinations of absolute values have been made by comparing the activities from degassed foils and from known amounts of active gas.

The same technique has been applied for measurements with Xe ions using the stable isotope Xe^{132} and the active Xe^{133}, with a half-life of about 5 days.

To measure relative Ne and Ar amounts in a target plate, we used the area under a (p, γ) resonance curve measured in a high voltage machine, and the absolute values were determined by means of a mass spectrometer.

The accuracy of the absolute determinations was about 5 per cent for Kr and Xe, and about 10 per cent for Ne and Ar.

2. Review of the Results and Discussion

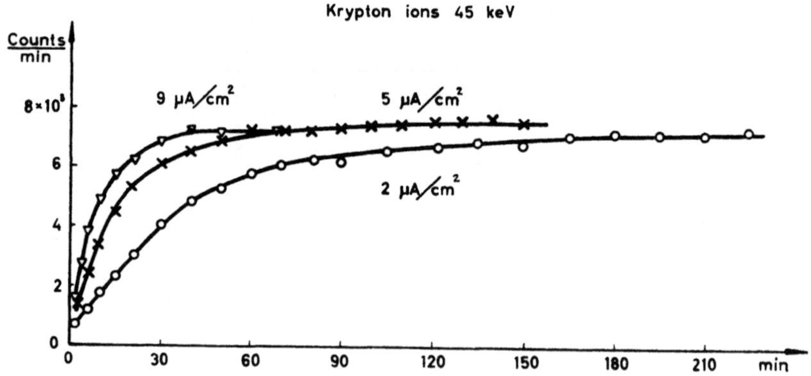

Fig. 2. Saturation curves in Al

Fig. 2 shows typical saturation curves for Kr ions on Al for three different current densities; they reach the same saturation value, here

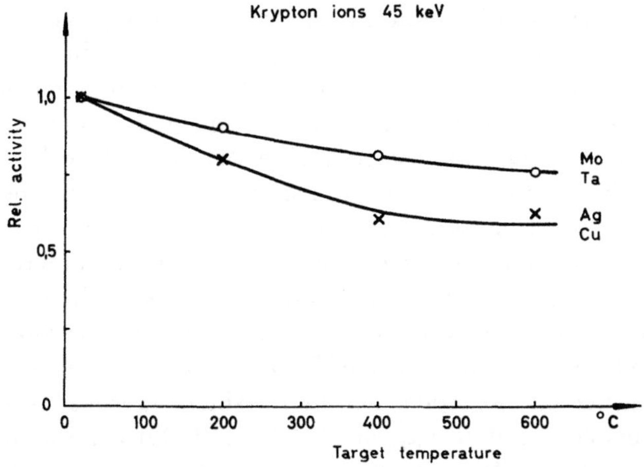

Fig. 3. Influence of target temperature on saturation values

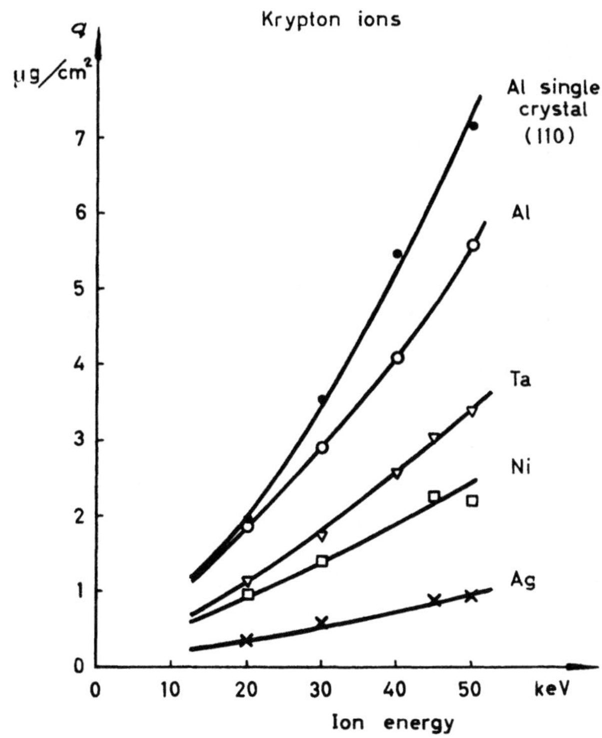

Fig. 4. Saturation quantity versus ion energy in some target materials

referred to as q, measured in $\mu g/cm^2$. However, the saturation value can decrease when the current density increases, due to warming up of the target. The influence of target temperature is illustrated in Fig. 3, where the measurements were made with constant current density and the target was cooled immediately after the collection was finished. The decrease was most pronounced for the low melting metals.

All subsequent measurements were made at room temperature. Fig. 4 shows the influence of ion energy and, also, that different target materials have quite different saturation values. Also the structure of the metal can influence the saturation value, e.g. the orientation of a single crystal, which is indicated in Fig. 4. In addition some preliminary measurements on Xe collection in Al single crystals of different orientations show that q (in $\mu g/cm^2$) is 2.9 for a (100) plane in the surface and 5.3 for a (110) plane, compared with 4.3 for polycrystalline Al.

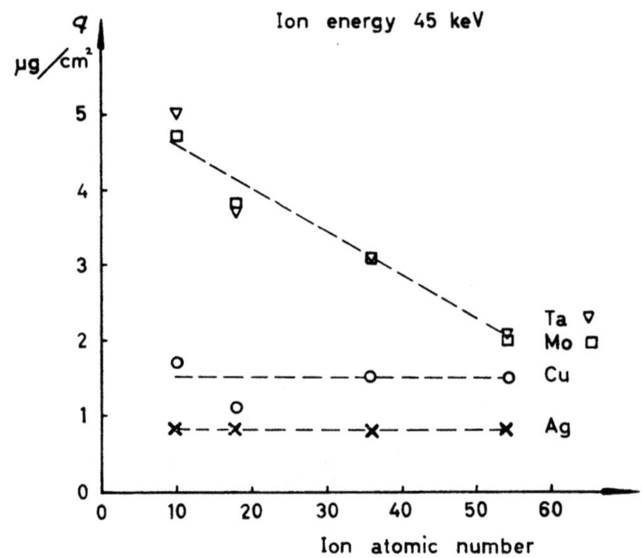

Fig. 5. Saturation quantities for different noble gas ions in some target materials

The saturation values for different noble gas ions are shown in Fig. 5. Here we see two types of ion mass dependence, which can be explained in the following way:

We define the fraction of trapped ions as

$$f = \frac{n_1}{n_1 + n_2}$$

where n_1 and n_2 are the number of atoms per unit volume. For small f values we can use the penetration depth d µg/cm², given by K. O. Nielsen [2], and we can approximately write the saturation quantity as

$$q \approx d\bar{f}\,\frac{M_1}{M_2}$$

where M_1 and M_2 are the mass numbers and $\bar{f}$ is the mean value within the depth d.

If we suppose $\bar{f}$ to vary slowly with the ion mass, the measured quantities in Ta and Mo agree with the formula, since a decrease in M_1 will give a still higher increase in d. In Cu and Ag and also in other soft metals there is a mass dependent decrease in $\bar{f}$, caused mainly by diffusion, which explains the other type of mass dependence [3]. Of course, there are target materials with ion mass dependence intermediate between the two cases above.

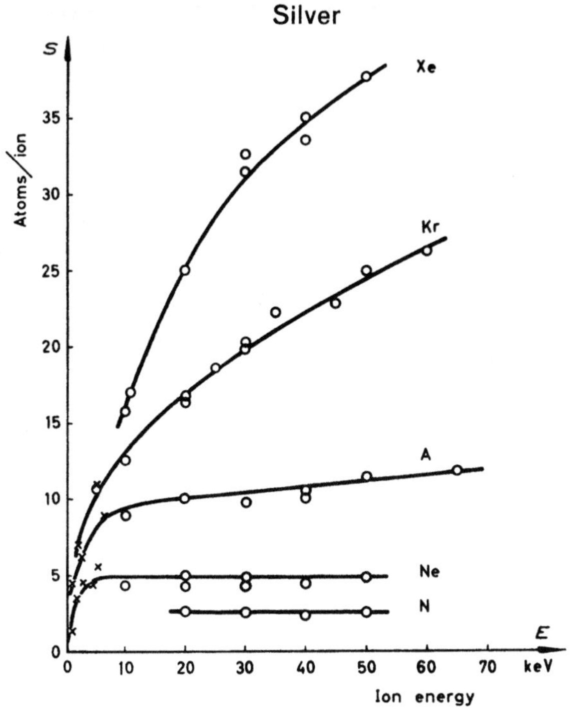

Fig. 6. Sputtering ratio of Ag for noble gas ions and N ions versus ion energy. Sputtering ratios from F. Keywell are indicated by ×

The collection mechanism also depends on the sputtering of the target material. Therefore, we measured sputtering ratios for noble gas ions bombarding different materials. The target was weighed before and after the bombardment by means of a micro balance. Removed quantities were a few hundred μg up to about 2 mg, on a target area less than 1 cm^2. The ion current was measured in a Faraday cage and the current density was of the order of 100 $\mu A/cm^2$.

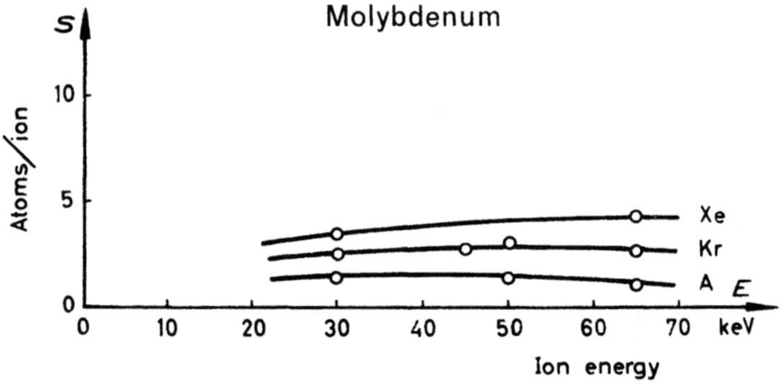

Fig. 7. Sputtering ratio of Mo for some noble gases versus ion energy

To give examples of the result, curves are shown for one high sputtering metal, Ag, in Fig. 6, and for one low sputtering metal, Mo, in Fig. 7. Experimental accuracy is about ±5 per cent.

Figs. 8 and 9 represent a summary of our collection and sputtering experiments on different materials using Kr and Xe ions of 45 keV energy. We see that there is a close relationship between sputtering and saturation value. The explanation is that both the diffusion coefficient and the sputtering ratio depend on the same fundamental properties of a material. We also observe a periodicity in the curves, familiar from the periodic table of elements.

References

1. Almén, O., G. Bruce, and A. Lundén, Nuclear Instruments **2**, 249 (1958).
2. Nielsen, K. O., Proc. 1955 Harwell Isotope Conf., Chapter 9.
3. Almén, O., and G. Bruce, Nuclear Instruments, in press.
4. Keywell, F., Phys. Rev. **97**, 1611 (1955).

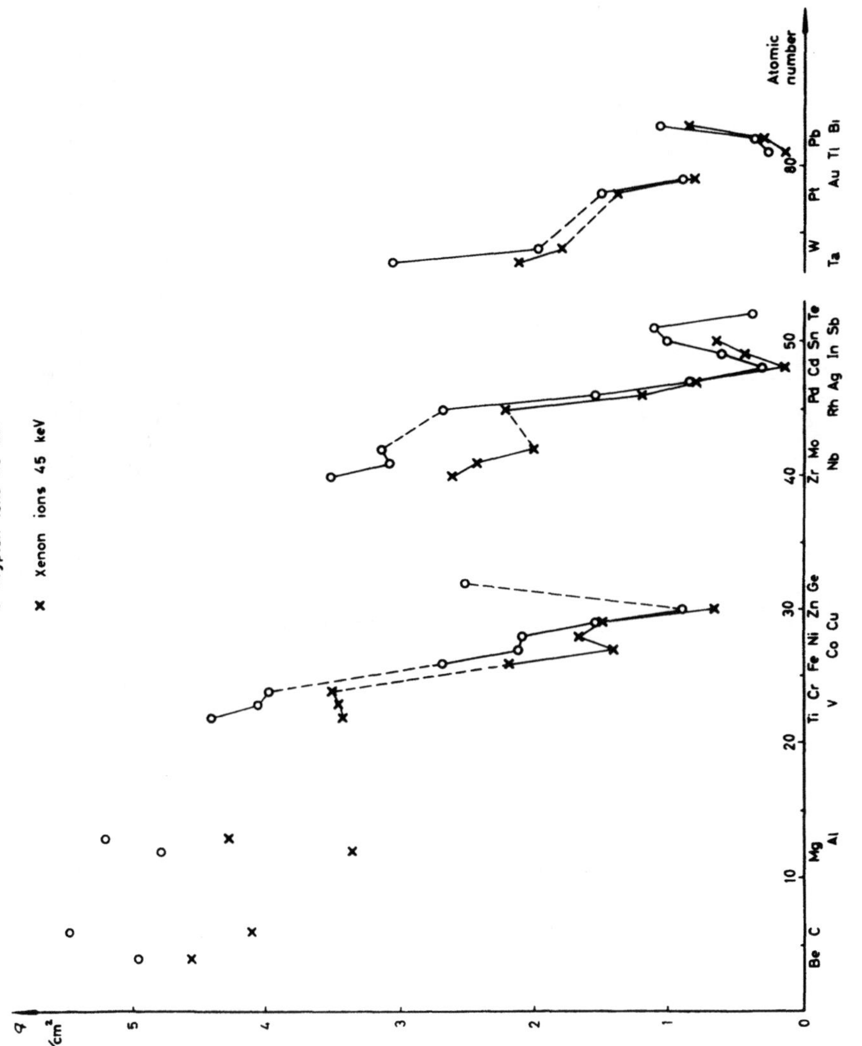

Fig. 8. Saturation quantities of Kr and Xe in different target materials

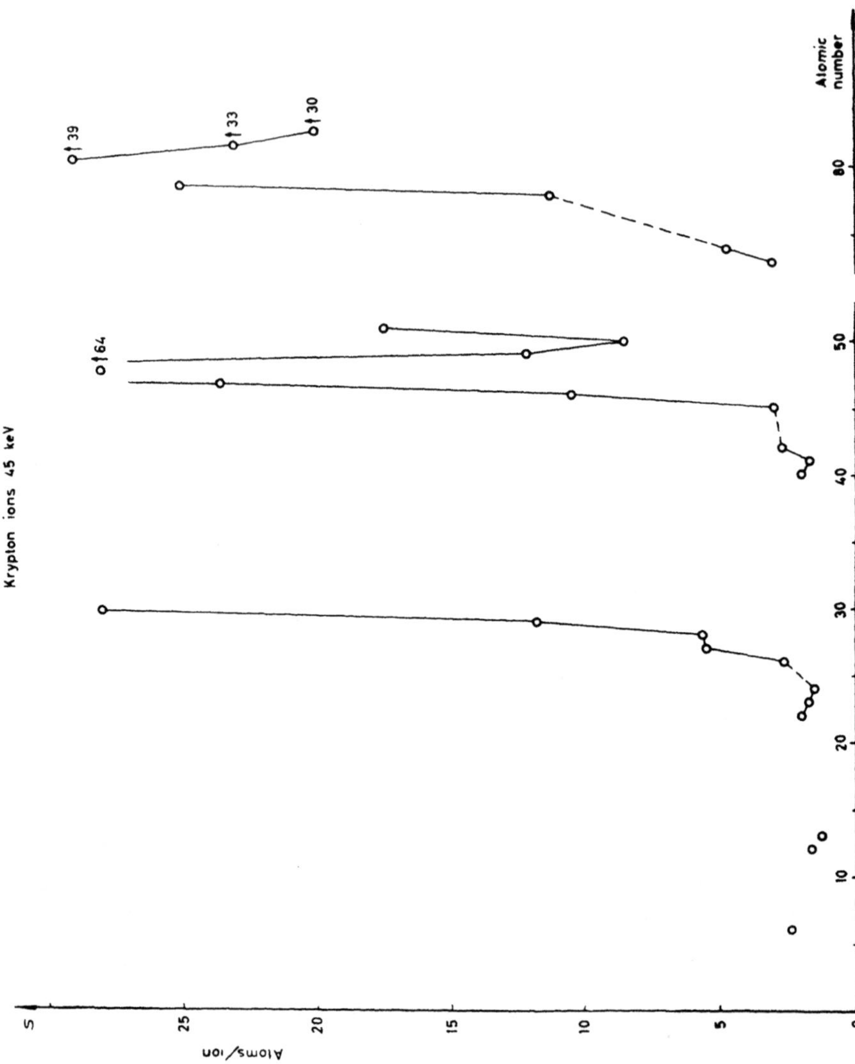

Fig. 9. Sputtering ratios of Kr ions on different materials. Xe sputtering is somewhat higher than that of Kr and has the same periodicity

Sputtering by Ions with 10 – 65 keV Energy

By

O. Almén and **G. Bruce**

Department of Physics, Chalmers University of Technology, Gothenburg

With 5 Figures

The sputtering experiments described in this paper were made with the 90° electromagnetic isotope separator in our laboratory [1].

1. Sputtering by Noble Gas Ions at Different Energies

Several authors have tried to calculate sputtering ratios from ion energy and mass numbers of ion and target atoms. The calculations according to P. K. Rol [2] seem to be in best agreement with our experiments. He used a fundamental work on the penetration of atomic particles through matter by N. Bohr [3] to obtain a formula for the energy dependence of sputtering ratios. The result is

$$S = K\pi R^2 n_0 \varepsilon E \quad \text{atoms per ion}$$

where $K =$ a constant, $R =$ collision diameter, $n_0 =$ number of atoms per unit volume, $\varepsilon = \frac{M_1 M_2}{(M_1 + M_2)^2}$ with M_1 and $M_2 =$ mass number of ion and target atoms, $E =$ ion energy.

We investigated the energy dependence of sputtering for more than 40 combinations of ion and target material. Typical sputtering curves are shown in our other paper[1] Figs. 6 and 7. Nearly all these curves can be well expressed by the equation above for ion energies between 10 and 65 keV, if K is given a suitable value for each curve. After some attempts to get an expression for K we found that $\ln K$ as a function of $\frac{E_B}{M_1 + M_2}$ gives straight lines in a diagram, one for

[1] See "Collection and Sputtering Experiments with Noble Gas Ions" by O. Almén and G. Bruce.

each noble gas (Fig. 1) E_B is the binding energy of the target atoms [4, 5]. The gradients of these lines are proportional to $-\sqrt{M_1}$ and we get

$$K = a \cdot e^{-\frac{b\sqrt{M_1}}{M_1+M_2} \cdot E_B}$$

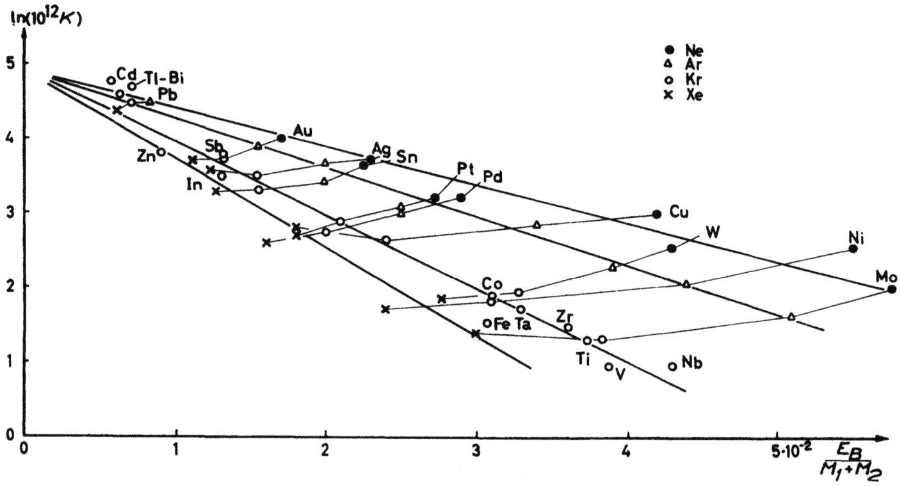

Fig. 1. ln K versus binding energy of target atoms divided by the sum of ion and target mass numbers. K is calculated from the sputtering equation and a number of sputtering curves

With numerical values for the constants a and b, we get the final equation for the sputtering ratio

$$S = 4{,}24 \cdot 10^{-10} e^{-10{,}4 \sqrt{\frac{\varepsilon}{M_2}} \cdot E_B} R^2 n_0 \varepsilon E \qquad \text{atoms per ion}$$

(E_B and E in eV, M_2 mass number, R in m, n_0 in m^{-3}).

The probable deviation between this formula and all our experimental results is about 20 per cent. The greatest deviations are found for Mg and Al, but on these metals a thick oxide layer decreases the sputtering.

2. Sputtering by Other Ions at 45 keV

Fig. 2 shows sputtering ratios for different ions of 45 keV energy impinging on three target materials: Cu, Ag, and Ta. We used these three because foils of good quality can be obtained commercially. The current density was 10–100 μA/cm² and the vacuum in the target chamber 0.8–1.5 · 10^{-5} mm Hg.

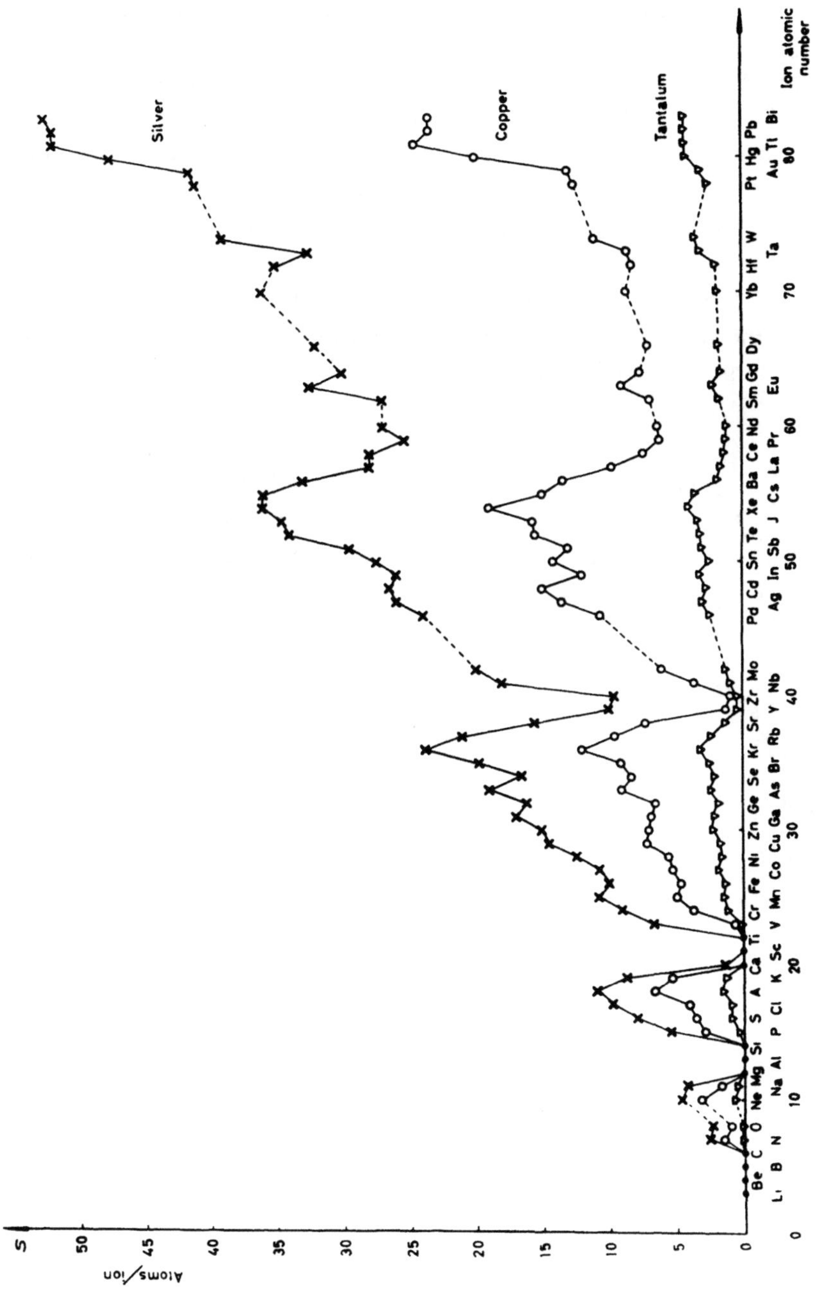

Fig. 2. Summary of sputtering ratios for different ions of 45 keV energy

Some regularities in the sputtering ratio can be observed in the diagram and will now be discussed.

a) Maximum sputtering ratios are noted for noble gas ions Ne, Ar, Kr, and Xe. The increase with the mass number is approximately proportional to the mass, but this is a chance occurence, since another choice of ion energy will give other curves.

b) Sputtering of the target material is sometimes zero. This means that the ions gradually build up a layer which protects the underlying material. The conditions for this are that

1. the self-sputtering, that is the sputtering ratio of ions on the same target material, is less than one,

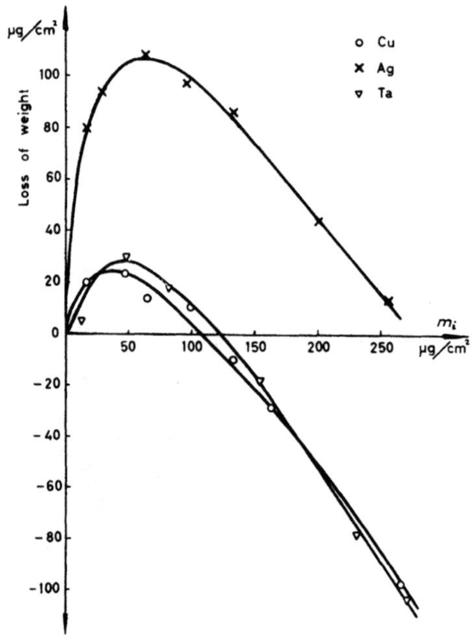

Fig. 3. Loss of weight of three different targets for increasing quantities of bombarding C ions

2. the initial sputtering of the target material is not too large,

3. the target temperature is sufficiently low to avoid evaporation of the deposit. Of course it takes a certain time for the ions to start building up layers and, therefore, a sputtering can be observed at the beginning of bombardment.

Fig. 3 shows the loss of weight of Cu, Ag, and Ta targets for increasing quantities of bombarding C ions. The ion beam is swept over the target in order to obtain a uniform current density. After about 50 μg/cm² of bombarding ions, the targets begin to increase their weights and the building up corresponds to a self-sputtering of about 0.4 atoms per ion. If bombardment is continued for a much longer time the deposited carbon layer may flake off because of the difference in thermal expansion of carbon and target material.

From Fig. 4 it is seen that building up of Ca ions takes place on Cu and Ta. On Ag the sputtering of the target material is too high to permit building up. A final sputtering of 1.5 is measured, which may be compared with the sputtering ratio 10.8 of equally heavy Ar ions on Ag.

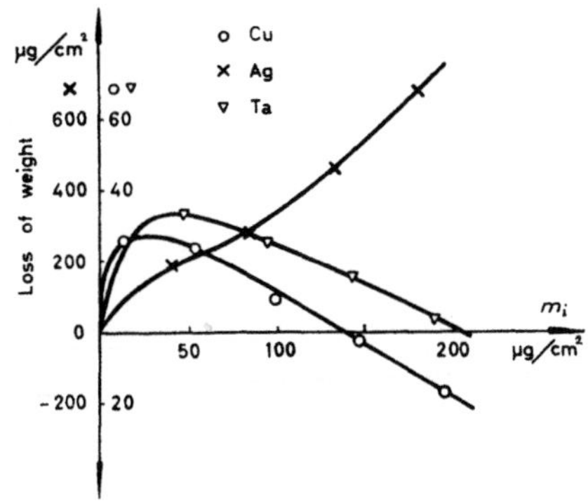

Fig. 4. Loss of weight of three different targets for increasing quantity of bombarding Ca ions

Building up is observed for Be, B, C, Al, Si, Sc, and Ti ions on the three investigated target materials and is expected on most materials, even at high current densities. Mg does not build up if target temperature is raised above 150–200° C due to condition 3 above.

Fig. 5 shows self-sputtering of some elements. We see that sputtering is below 1 for all ions building up and above 1 for all other ions.

Sputtering is about zero for very light ions. We have used He and Li ions on Ag and measured sputtering ratios less than 0.1 atoms per ion, but we have not investigated this further.

c) Most ions have sputtering ratios between the ratios for noble gas ions for corresponding mass numbers and zero. We know saturation quantities for noble gas ions on different materials. They are of the order of a few $\mu g/cm^2$. We have also measured saturation for Rb ions and found it somewhat higher than that for Kr. When sputtering decreases to zero, the collected quantity may be infinite. Therefore, most ions can be collected in quantities of at least a few $\mu g/cm^2$ with the direct collection method and the final sputtering ratio indicates how much larger the saturation quantity is compared with that of a corresponding noble gas.

d) Finally it may be noted that also related elements other than the noble gases appear in similar positions on the sputtering curves. E.g. we point out the regular decrease for the ions immediately after the noble gases. In minimum positions we find Sc-Ti, Y-Zr, and the La series—Hf.

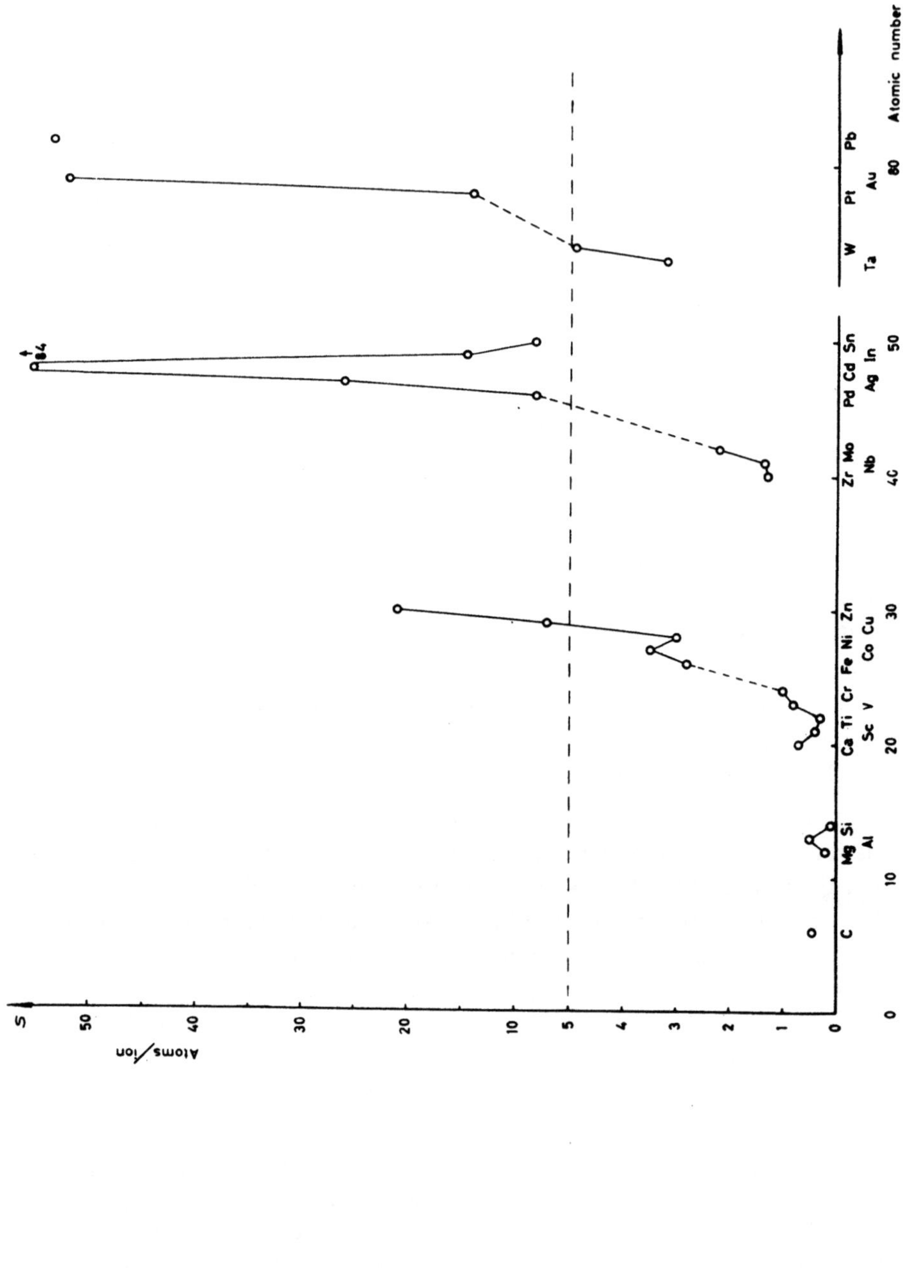

Fig. 5. Self-sputtering ratios for some ions of 45 keV energy

Our results can be compared with reports in literature. In Proceedings of the Harwell Conference M. L. SMITH [6] pointed out that B, Si, and Ti build up layers. This is in agreement with our results. However, P. K. ROL [2] measured a sputtering ratio of about 4 for 20 keV Si ions on Cu. In the Conference report it is also said that Sn and W build up. But we cannot understand that at this moment. In some cases we know that sputtering decreases somewhat with increased target temperature. Therefore, we investigated temperature dependence of Sn sputtering on Cu but found it independent of temperature below 500° C. In addition, the self-sputtering of Sn and W is above 1, see Fig. 5.

Sputtering ratios on Cu reported by P. K. ROL [2] are mostly in agreement with our results, while the results of O. C. YONTS and al. [7] are about 25 per cent higher, which may be due to their much higher current densities.

References

1. ALMÉN, O., G. BRUCE, and A. LUNDÉN, Nuclear Instruments **2**, 249 (1958).
2. ROL, P. K., Thesis (Amsterdam 1960).
3. BOHR, N., Mat. Fys. Medd. Dan. Vid. Selsk. 18, no. 8 (1948).
4. BREWER, L., in: Chemistry and Metallurgy of Miscellaneous Materials Nat. Nucl. Energy Series Div. IV. vol. 19 B, p. 13.
5. Thermodynamic Properties of Metals and Alloys, Min. Res. Lab. University of California.
6. SMITH, M. L., Proc. 1955 Harwell Isotope Conf., Chapter 7.
7. YONTS, O. C., and al., J. Appl. Phys. **31**, 447 (1960).

Preparation of Targets and Radioactive Sources in an E.M. Separator

By

G. Sidenius and **O. Skilbreid**

Institute for Theoretical Physics, University of Copenhagen

With 5 Figures

In isotope separators of "medium size" the two methods most extensively used for the collection of isotopes are direct deposition of the ions and collection by sputtering.

The direct deposition method is almost exclusively used for the preparation of radioactive sources for β- and γ-spectroscopy. In these cases the radioactive ions are usually collected on aluminium foils with a thickness of 150 μg/cm².

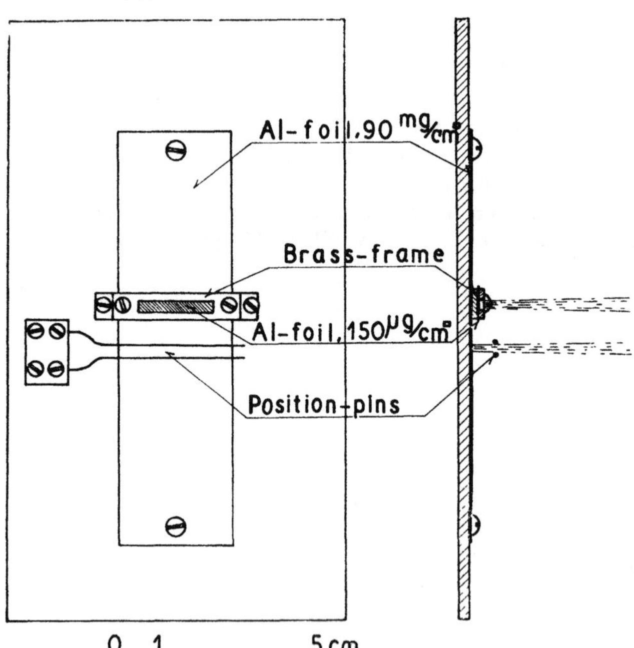

Fig. 1

In Fig. 1 the collecting device used in the preparation of β-sources is shown. The thin foil is mounted in a frame with a 20 × 3 mm² slit.

Silicon grease is used to stick the foil to the backing and to protect it against heat destruction, in cases where there are stable isomers or isobars originating from the carrier or from impurities in the charge material. With the foil mounted in this way, beam currents of the order of 10 μA can be tolerated without destruction of the foil.

However, this method suffers from the disadvantage that the penetration of the ions into the surface puts a lower limit on the effective layer thickness obtainable for the material being collected. The direct deposition method also suffers from the disadvantage that it only permits a limited amount of material to be collected for most elements. This is due to sputtering out of the collector, which for most elements causes an equilibrium to be set up between the material being deposited and that already collected. Furthermore, only a very limited beam current can be used if the deposition is to be made on thin foils.

The sputtering method has the advantage that the amount of material that can be collected is substantially greater, and because the sputtered atoms have low energies they can be collected on thin foils. Therefore this method has been used with success for the preparation of a great number of stable targets for the Van de Graaff group. Most of these separations have been made with pure reactor graphite as sputtering backing and with carbon film of about 100 $\mu g/cm^2$ as collecting foil. The reason why we chose graphite as sputter backing was, firstly, that we did not add new impurities in the target. Secondly, we have found by experience that the ratio between the desired material sputtered and the backing material sputtered is very favorable.

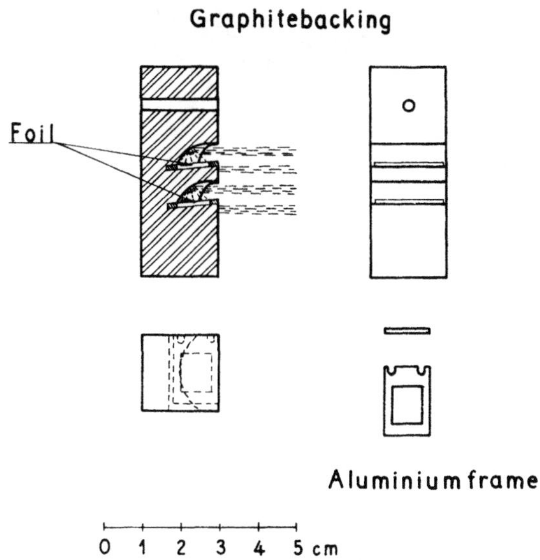

Fig. 2

The collecting device used is shown in Fig. 2. The separated beams impinge on a curved surface milled in the graphite block. In the

C direction the centre of the radius is on the foil. A considerable fraction of the atoms is sputtered from the surface and caught on the thin carbon foil which is mounted on a frame of aluminium. The foil faces the beam spot on the graphite, and is placed in such a way that it is protected against the direct beam. For mass numbers greater than 130 the mass dispersion and the dimensions of the slits only allow either the even or the odd isotopes of an element to be collected simultaneously. For a 20% abundant isotope it normally takes about two hours to prepare a target of ideal thickness (about 30 $\mu g/cm^2$).

In one case, the sputtering method was also used successfully in the preparation of a radioactive source. The purpose was to collect Th^{232} directly on a foil of plastic scintillator of thickness 14 mg/cm^2 and with an area of about 3 cm^2. The collecting device is shown in Fig. 3. Graphite is used as sputterer backing. To obtain a plane sur-

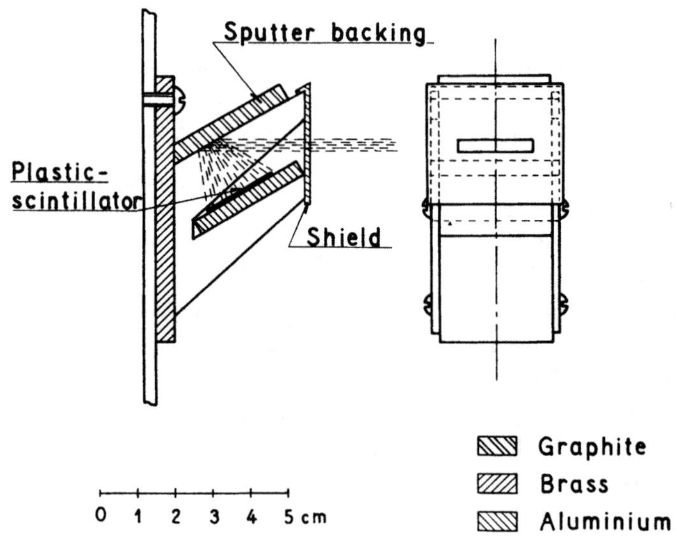

Fig. 3

face of collected material on the foil, the distance between the sputtering surface and the collecting foil is made rather great. The beam current was about 60 μA and the collecting time was about 20 hours.

As a collector for radioactive isotopes, the sputtering method suffers from the disadvantage that the collected sample is mixed with material from the sputterer backing and that the collected material is spread over a great area.

By retarding the ions before collecting them we have been successful in overcoming some of the difficulties mentioned for the above methods. By this technique, which was first used by OLIPHANT and YATES, the energy of the ions to be collected is reduced by means of an electric potential from about 50 keV to several hundred volts. It is then possible to collect the ions on thin Formwar-foils strengthened by a layer of evaporated aluminium or on thin aluminium foils. Furthermore, the material is concentrated on a very small area and the penetration of the particles into the surface of the backing is prevented.

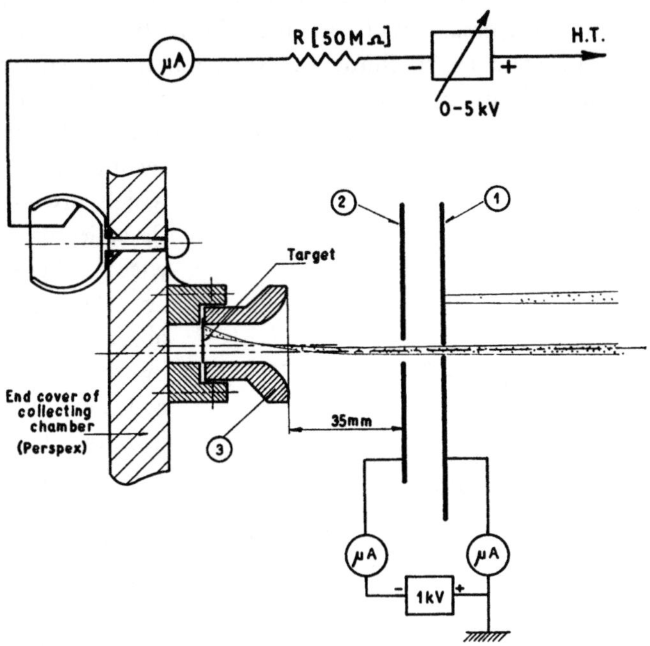

Fig. 4

The setup is shown in Fig. 4. The ions to be collected pass a slit in a base plate (1). Secondary electrons are prevented from reaching the collector by a guard plate (2) at a negative potential of about 1 kV. The energy of the ions impinging on the foil is adjusted by means of a variable voltage inserted between the resistor *R* and the high voltage of the separator. In order to reduce the electrostatic forces on the foil and to yield a fairly strong focusing effect, the collector electrode (3) is given the shape shown in the figure. This focusing is actually so strong that the focal point lies in front of the foil. The deposition takes place within an area of approximately 1×5 mm^2. It was found preferable to let the beam enter the collector a little off centre in order to utilize the energy separation of the ions in the electrostatic field. While no investigation was made to find whether this really gives an increase in enrichment, it was found that in fact the deposition occurs in two separate spots, the lower probably being caused by neutral atoms.

If the voltage difference between the ion source and the collector is decreased to below 1 kV, the amount of material collected per unit time is found to decrease in the manner shown by the dotted curve in Fig. 5. The solid curve shows the variation in the current from the guard plate. This current is caused by ions which do not reach the collector, but return to the guard plate. The returned ions release secondary electrons which are caught by the collector. The presence of these secondary electrons can be observed by measuring the current to the collector. This current is found to decrease and to reverse direction when the voltage V_{dif} is decreased. Simultaneously an X-radiation from the collector device can be measured.

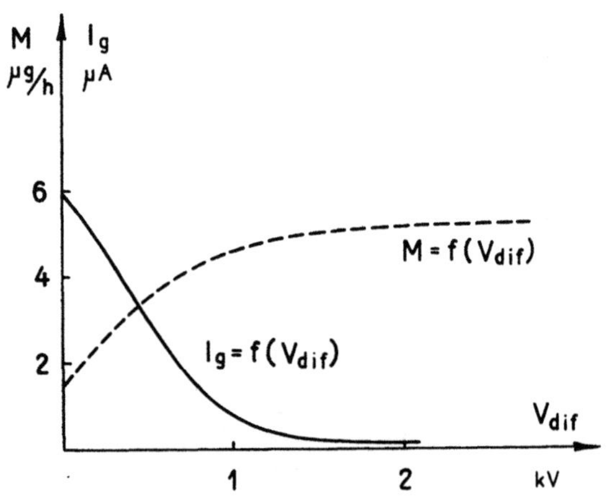

Fig. 5

By using a grid instead of the plate the number of electrons released by the returned ions can be reduced.

The energy spread between the great majority of the ions is less than 50 V. Therefore, we expect the phenomena described to be caused mainly by the space charge. Rough calculations show that the space charge should be expected to set up a potential barrier of the order of several hundred volts at beam currents of 1 μA/mm^2 and a voltage difference of 1 kV.

A confirmation of our interpretation is the fact that the voltage difference at which ions begin to return increases when the beam current increases.

Until now the method has only been used for the preparation of stable targets (1). However, using this method, highly concentrated sources can be prepared for β-spectroscopy. The main drawback of this method is that only one isotope of an element can be collected per run, but, especially in the preparation of β-sources, this drawback will be without importance.

Discussion

J. H. FREEMAN: When preparing targets by sputtering it should be mentioned that the sputtering distribution is not always normal to the sputtering surface. In the sputtering of bismuth on carbon the bismuth tends to sputter back along the beam direction even for very small angle incidence. This is probably simply a function of the imperfections of the carbon surface but it may upset the distribution on the collected targets.

P. K. ROL: We have also observed that for oblique incidence of magnesium ions on copper.

R. BERNAS: Did you analyse the composition of the two spots you obtained after retardation?

O. SKILBREID: We did not really make an analysis, we just made measurements of the relative intensity between these two deposits. We made it with the heavy particle spectrometer in connection with the Van de Graff machine. We found that the spot caused by neutral atoms had an intensity of about 10^{-4} of the spot of collected ions. We found that the neutral atoms were able to destroy the foil and that is the reason why we use energy separation.

R. BERNAS: What is the highest current you can use with this method?

O. SKILBREID: The beam current could be increased but the method was originally developed for the collection of very rare isotopes; it was developed for the collection of Lu^{176}. But if you make your collector in such a way that you don't get strong focusing you can achieve higher currents. One of the reasons why we wanted to concentrate the beam so much was to decrease the separation time. The Van de Graff group only needed a spot of 0.3×3 mm.

R. BERNAS: You mentioned an energy spread in the ion beam at the collector of the order of 15 eV. On what did you base this value?

G. SIDENIUS: We based it on the line shape of the normal spectra.

O. ALMÉN: I would like to point out the possibility of using this very good method also for the collection of gas isotopes. One would expect that it should be impossible to collect gas ions if the ion energy is too low. But in fact one can go down to about 5 keV and still get approximately the same ratio between trapped ions and target atoms in a thin surface layer. We have used retarded gas targets for (p, γ) resonance studies and in this way got a considerably increased resolution.

Sputtering at High Energy*

By

G. Ranc

C. E. N. Saclay, France

The paper concerns some developments of an observed deviation from constancy for the sputtering ratio when a copper target is bombarded by a cadmium ion beam and when the density of this beam is varied. All parameters being kept constant, the sputtering ratio is higher than unity when the density is high and smaller than unity when it is small (building up of cadmium).

A two-step mechanism of sputtering is then suggested. The impinging ions begin by damaging the surface structure of the target and atoms can be expelled only when the surface is damaged. There is an equilibrium between the creation of the damage and the restoration of the structure under thermal motion.

It was attempted to get corollary information by varying, not the rate of damaging, as formerly, but the rate of restoration by varying the temperature of the target. When irradiating a copper target with a cadmium ion beam (of constant density and energy) it was found after a few trials that, when a temperature gradient was imposed across it, the sputtering ratio was higher than unity at room temperature (the cold end of the target is sputtered), whereas the ratio is smaller than unity at the warmer end (200° C), cadmium builds up. The density of the ion beam is about 70 μA/cm^2, the energy 30 keV. The incidence was normal.

Much work remains to be done in this field, but at the present poorly elaborated stage, the above mentioned explanation can be kept in mind.

Discussion

O. Almén: How did you establish the build up of cadmium ions?

Ch. Cassignol: On the sputtered part of the copper target the surface was reddish and rough, on the part where cadmium was built up the deposit was whitish. It was only a qualitative test but I do not believe there can be any gross mistake.

---

* Ref. Cassignol and Ranc, Comptes rendus, **248**, 1988 (1959).

O. Almén: It is our opinion that the colour alone does not ensure a build up because even minute quantities may change the target colour completely.

Ch. Cassignol: In the one case there was a rough surface, in the other it was nicely polished.

M. J. Higatsberger: Except in Dr. Ranc's paper it seems that people working in the field of sputtering phenomena do not attach too much importance to the target temperature. In order to get a deeper and more coherent understanding of the effect, I think it is necessary to work down to target temperatures close to absolute zero with clean monocrystals.

Vacuum problems are not mentioned either in the former papers but we must realize that monomolecular layers are built up in seconds when the pressure is of the order of 10^{-5} mm Hg. Because the sputtering experiments were done at pressures of 10^{-6} mm Hg and the observation times were 100 sec. and longer more than one monomolecular layer had time to be built up.

L. O. Love: We had some trouble in collecting large Li ion currents. The retention was in the order of 20% and the copper collectors were failing in 24 hours. In effect, we were destroying the copper collectors instead of collecting Li. Li and Cu have an amalgamating property that permitted good thermal contact between the deposited Li ions and the copper collector. By cooling the collector to approximately 5° C, we were able to increase the beam intensity to approximately 1 ampere. The retention went up to 95% and the collector life was increased to 300 hours. In our experience, cooling the collector both protects it and retains the incoming isotopes. In this respect cooling has also proved beneficial when graphite is used for collection with certain other elements.

Chapter VI

Chemical Aspects in E. M. Separation Processes

E.M. Separations with High Efficiency of Microgramme Quantities

By

G. Sidenius and **O. Skilbreid**

Institute for Theoretical Physics, University of Copenhagen

With 3 Figures

In the E.M. separation of rare earths it is necessary to convert the commonly occurring rare earth oxides into the more volatile anhydrous chloride form. The material is then transferred to the ion source oven, from which the chloride is evaporated into the ion source. The speed of evaporating is controlled by controlling the oven temperature. This method has proved to be very useful when the amount of material to be separated exceeds 50 mg.

The ion source is of the oscillating electron type which has been described earlier elsewhere. The discharge chamber is made of pure graphite while the insulators are made of SiO_2.

If the amount of material to be separated is too small, the method mentioned becomes inapplicable for the following reasons. Firstly, small amounts of rare earth oxides are difficult to convert in a controllable way. Secondly, rare earth chlorides normally adhere strongly to the walls of the container in which they are made. Therefore, the material is difficult to transfer to the oven with a reasonably high efficiency. In addition, appreciable losses will occur because a large fraction of the very hygroscopic rare earth chloride will pick up water vapour from the air and convert into the inevaporable hydride form during the transferring process. Finally, for reasons dealt with elsewhere, it is difficult to obtain a constant rate of evaporation and hence stable running conditions if the amount of material to be evaporated is below a few milligrammes.

The various difficulties mentioned might be successfully overcome by adding a suitable amount of carrier to the material which is to be separated.

This, however, will increase the running time in direct proportion to the amount of carrier added and will be undesirable, particularly when short-lived activities are going to be studied.

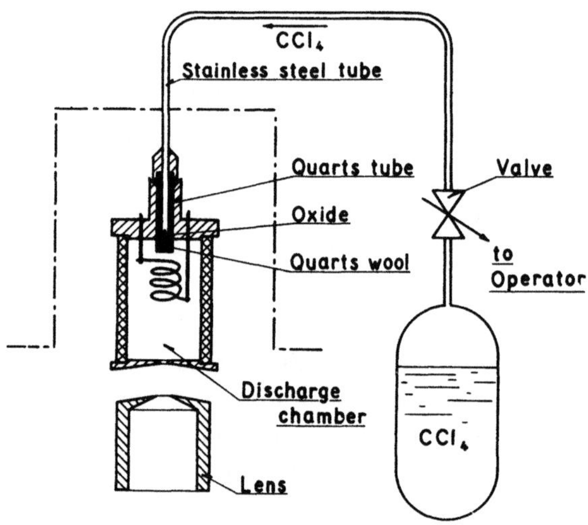

Fig. 1

We have found that reproducible separations of microgram amounts can be obtained with high efficiency by the following method which is shown in Fig. 1. The material to be separated in the form of oxide is placed in a filter of quartz wool which first has been placed into one end of a quartz tube. The other end of the quartz tube is connected to a tube of stainless steel. By means of a vacuum lock the tube is inserted in the isotope separator and connected to the ion source in such a way that the free end containing the oxide sticks about 5 mm into the discharge chamber. The gas inlet tube is via a needle valve connected to a container with CCl_4. As soon as the tube has been placed in the position shown, CCl_4 is fed into it. The rate of flow of CCl_4 is kept sufficiently high to maintain a stable plasma in the ion source. After a few minutes, when the tube has reached a sufficiently high temperature, the rare earth oxide converts into chloride. The chloride now feeds directly into the discharge chamber.

Fig. 2 shows in detail the connection of the quartz tube to the gas inlet tube. The gas inlet tube and the two parts of the connection

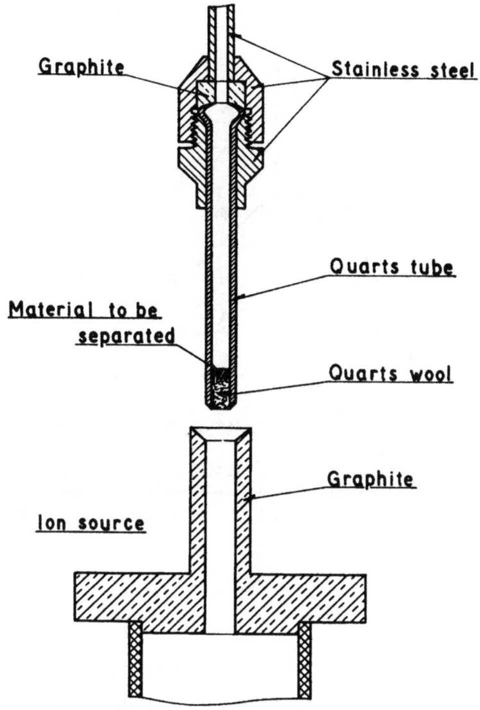

Fig. 2

union are made of stainless steel. A graphite packing is used for sealing the connection.

Fig. 3 shows the current of the rare earth as a function of the CCl_4 pressure. The pressure was measured by means of a Penning gauge which was connected to the main vacuum system. The total ion current from the ion source was approximately constant during the measuring period. We assume that the current I_R of the rare earth is proportional to the relative pressure of the rare earth chloride in the ion source. If this is so, a function proportional to the total pressure of rare earth chloride in the ion source and hence the rate of conversion can be found by multiplying I_R by the total pressure of CCl_4. The latter is an order of magnitude higher than the pressure of the rare earth chloride. The curve obtained is shown dotted in the same figure.

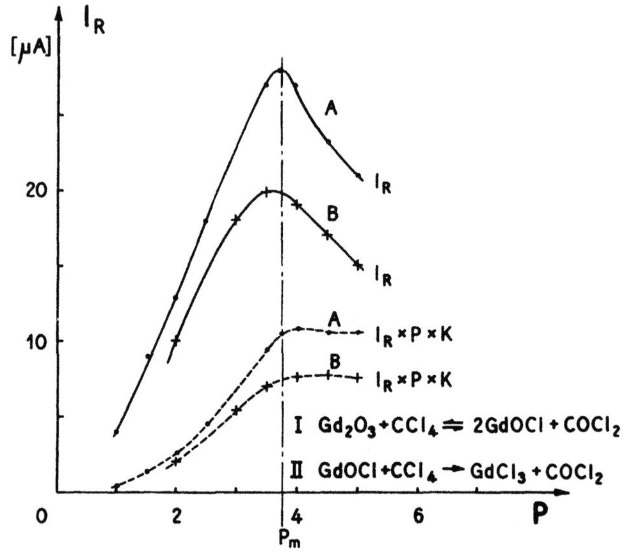

Fig. 3

These curves indicate that the rate of conversion is approximately constant as soon as the CCl_4 pressure exceeds a certain value Pm. This might be so because most of the surface of the oxide takes part in the converting process at pressures above Pm, and a saturation effect is obtained. This assumption is consistent with our experience that the rare earth ion current obtainable is roughly proportional to the amount of oxide present in the tube. At pressures below Pm, the curves indicate that the relative increase in the rate of conversion is higher than the relative increase in the pressure of CCl_4. We have discussed this phenomenon with members of the chemistry group at this institute. A possible explanation might be that the formation of the rare earth chloride, for instance, takes place in a way suggested by the two equations in Fig. 3[1]. If the reaction given in the first equation is reversible, the amount of, in this case, GdOCl will be roughly pro-

[1] These e. g. only serve as an illustration and are not believed to describe the process really taking place.

portional to the pressure of CCl_4. If this guess has anything to do with reality, we should expect the rate of conversion to be proportional to the second power of the CCl_4 pressure. In order to make an approach to the solution of these questions, we have decided to make more direct and accurate measurements of the rate of conversion at different CCl_4 pressures. In these experiments we plan to use some radioactive rare earth oxide, measure the pressure of CCl_4 in the quartz tube, and collect the converted material on a metal disk.

If we adjust the pressure of CCl_4 so that maximum current of the actual rare earth is obtained, we can carry out a separation in approximately one hour. This duration is largely independent of the amount of material to be separated. If we are correct in our guess that only a limited rate of conversion is obtainable with a given surface area of the oxide, we achieve in this way the smallest possible separation time. In most of the separations carried out by this method, the ion source was operated at conditions which allowed us to obtain the above mentioned maximum current. However, these conditions are not necessarily those that give maximum efficiency. According to investigations made by ALMÉN and K. O. NIELSEN with ion sources of the type we use, the efficiency is about inversely proportional to the pressure in the discharge chamber. Therefore, if the highest possible efficiency is wanted, the pressure of CCl_4 should be adjusted so that the pressure in the discharge chamber is just above the minimum working pressure for the ion source. In order to obtain both high current and high efficiency, an adjustable flow resistance should be present between the quartz tube and the discharge chamber so that the pressure in the quartz tube and the ion source could be adjusted independently. Hitherto a fixed flow resistance consisting of a plug of quartz wool has been used. For most of the separations we have made by this method, the efficiency has been in the region of 5% to 15%. In a single case an efficiency of 22% was measured. We believe that the changes in efficiency from one separation to another are due to changes in the above-mentioned flow resistance at the outlet of the quartz tube. However, the efficiencies we obtain by this method are appreciably higher than those obtained when a rare earth chloride is evaporated from an oven. In the latter case, where the efficiency is 1–2%, a large fraction of the chloride is probably lost by decomposition in the ion source. By the new method, these losses are believed to be much less because the great surplus of carbon tetrachloride present in the ion source reconverts the decomposed rare earth chloride.

In Table 1 the elements which have been separated by this method are listed.

Table 1

Appl.					
Element	V.D.G.	β	γ	l.t.	Activity produced by
Sm	—	—	—		Pile
Eu	—	—	—		Cyclotron
Gd	—				
*Tb		—	—	—	C.
Dy	—				
Ho		—	—		C.
Er	—				
Tm		—	—	—	C.
Yb	—				
Hf	—	—	—		P.
W	—				
Th	—			—	Natural
Pa		—	—		C.
U	—				
Pu				—	P.

Discussion

K. O. NIELSEN: Dr. SKILBREID mentioned that in order to get high efficiency the pressure of the supporting gas should be low. I believe it would be possible to gain fairly high efficiency by a proper mixture of supporting gases.

O. SKILBREID: It may be that one can obtain higher efficiencies by a mixture of two different gases in the ion source. We found, however, that with a high pressure of CCl_4 in the source we got the lowest efficiency.

J. H. FREEMAN: I would just like to suggest that you try phosgene instead of CCl_4. We find that phosgene is an extremely efficient chlorinating agent, it is a gas and the pressure can easily be controlled.

O. SKILBREID: We have not tried phosgene but we have tried pure chlorine, chloroform and freon and we have actually not found much difference between these gases. Of course, chlorine was quite different from the other chlorides.

P. WEINZIERL: Have any difficulties been experienced in irradiating the oven of an ion source in the reactor? The compound of the source material could be influenced by some Szillard-Chalmers effect during the (n, γ) reaction.

O. Skilbreid: When using chlorides made before irradiation we have not really found any noticeable decrease of the efficiency because of decomposition in the pile at a flux of 10^{12} neutrons/cm^2/sec. With the higher flux (10^{14} neutrons/cm^2/sec) we now have available we use oxides and convert them afterwards directly in the ion source as described in the paper. But we do not know if there is any difference between the efficiency of the stable and the radioactive atoms.

Some Chemical Problems Connected with the Electro-Magnetic Separation of Isotopes

By

A. E. de Vries

F.O.M.-Laboratorium voor Massaspectrografie, Amsterdam

The Amsterdam isotope separator has been used for several years and during that period we have learned quite a few things which were of importance, some of them rather unexpected. In this paper some of these problems will be discussed.

Source Material

a) Choice of source material

In the early separations we always used gases or vapors as source materials. A simple consideration teaches, and this was confirmed by the separations carried out later on, that a material with a low pressure at room temperature is preferable.

Ions are lost in the vacuum chamber by processes which may be described by a free path λ. The current at the collector is then given by

$$i_c = i_s \exp -l/\lambda$$

in which i_c = collector current, i_s = current from the source, l = length of vacuum chamber.

In general the current from the source of an electromagnetic separator is within reasonable limits independent of the pressure inside the source [1]. λ is inversely proportional to the pressure p_v in the vacuum chamber, so we can write

$$i_c = k \exp - C l p_v$$

in which k and C are constants. This means that the current increases with decreasing pressure. Of course, enough gas has to be present to compensate the space-charge, but this is the case at a much lower pressure than under normal working conditions. In our separator the beam intensity is still increasing with decreasing pressure in the neighbourhood of 10^{-6} mm Hg.

Now if one uses materials which have to be heated to get the right pressure in the source the material will not increase the pressure in the vacuum chamber which is at room temperature. In this case we usually get ion currents which are 5–10 times as large as with gases.

For lower ion currents, in which case the current from the source is proportional to the pressure in the discharge chamber, there appears to be an optimal pressure given by the condition that λ is equal to the length of the ion path in the vacuum chamber [2]. Also in this case the pressure in the vacuum chamber would be too high for volatile materials.

A second reason for using low-pressure substances is the fact that vapour cannot contaminate the collector.

If it is possible to use the element this is advantageous. Drawbacks of compounds are:

1. the formation of other ions and therefore in general a lower current of the desired ions, and

2. the possible formation of compounds or elements with a high vapour pressure.

For instance, with chlorides, enough chlorine is produced to increase the pressure and thus decrease the ion current. This is not completely suppressed by the use of cold baffles.

b) Purity of the source material

In general we found that impurities in the source material can be very troublesome. Separations of the uranium isotopes were carried out with UCl_4 made by leading chlorine over uranium. In this reaction, which has to be started by heating the uranium but then proceeds by itself, other chlorides are formed as well. One of them, UCl_6, has a rather high vapour pressure at room temperature and therefore can be removed better. UCl_3 has a very low vapour pressure even at temperatures where one uses the UCl_4. We always found a considerable quantity of the source material sintered together in the oven after the separations until we purified the UCl_4. Then all the material could be used in one run.

Probably one of the reasons for the difficulties is the lowering of melting range and the subsequent forming of a barrier for the vapour.

Collector Material

a) The rate at which the separated material leaves the collector is given by

$$Q_1 = 5.8 \cdot 10^{-2} Ap\,(TM)^{-1/2}\ \text{Moles/sec} \qquad [2, 3]$$

in which A is the area of the collector slit in cm^2, p and M are the vapor pressure and the molecular weight of the material in mm Hg and T is the temperature in °K.

The rate of supply to the collector is found from the electrochemical equivalent:

a current of 1 Amp. is equal to 1/96,500 Moles per second.

If the abundance of the desired isotope is $a\%$ the supply is

$$Q_s = a/100 \cdot 1/96{,}500\ \text{Moles}$$

and if the current is i_c mA we get

$$Q_s = a/100 \cdot 1/96{,}500 \cdot 10^{-3} \cdot i_c$$

$$Q_s = 1.04 \cdot 10^{-10} a i_c\ \text{Moles/sec}$$

If the loss has to be less than 1% of the supply one gets

$$p < \frac{1.8 \cdot 10^{-11} a i_c (TM)^{1/2}}{A}\ \text{mm Hg}$$

For $a = 1\%$, $i_c = 1$ mA, $T = 300°$K, $M = 40$ and $A = 20\ cm^2$ we get the condition

$$p < 10^{-10}\ \text{mm Hg}$$

This means that quite a few substances which have a 'low vapour pressure' in the ordinary sense of the word, like sulphur, have to be chemically bound at the collector. Sulphur, for instance, is trapped on a collector with copper turnings and forms non-volatile copper sulphide.

b) It goes without saying that in the choice of the collector it is very important to look for impurities of the element which is to be separated. In most cases we used ordinary solder which contained quite a percentage of cadmium in separations of the latter element. The choice of the collector is also important in connection with the processing of the material. There are cases where a simple burning of a graphite collector pocket leaves behind the separated isotopes. In this case a collector of copper or aluminium might cause a considerable increase in the labour involved in the processing.

Processing of the Separated Material

Before the separations started we thought that the separated isotopes would be deposited on the collector as a layer of more or less pure material. That this was not so we found by chance. To determine the most favourable form of the collector we carried out short runs of half an hour or so and then analysed the amount of material trapped on the collector. While a colorimetric analysis of one of the first runs in which silicon was separated on an aluminium collector indicated an efficiency of $\sim 20\%$, an estimation by Newton rings indicated almost complete trapping of all ions on the collector.

(A certain wave-length will be extinguished if $2d/\lambda + 1/2 = k$, in which d is the thickness of the layer, k is an integer and λ is the wave-length in the material. For the thinnest layers this leads to the following colours:

thickness of layer $\times 10^5$	2	4	5	6	7	cm
colour	grey-blue	yellow	red	purple	violet)	

When the same colours appeared with a separation in which a rare gas was used it was clear that a bigger part of the layer was formed by the material of the collector. The sputtering coefficient in the first experiment thus appears to be about 4 (four aluminium atoms sputter together with one silicon atom to the opposite side of the collector).

The fact that alloys and compounds may be formed is one of the reasons why chemical properties of separated elements are not always the same as under normal circumstances. The crystal structure may be quite different too. A third cause probably is the fact that impurities tend to be of the same mass rather than of the same chemical behaviour. It is well known that extremely pure substances differ in their properties from normal materials. A few examples will be given now.

Although uranium is a very base element, no material could be detected after scraping of the graphite surface of the collector and subsequent treatment with acids. This was found in a short run, while for longer separations uranium could be obtained by direct leaching. In that case the procedure, of course, is not quantitative. The best and easiest method is to burn the graphite completely and then dissolve the uranium for further purification.

A similar observation was made with zinc separated on a copper collector. In this case not only the formation of an alloy forced us to dissolve the copper as well, but also part of the zinc which was present as the element could not be dissolved in hydrochloric acid, which is certainly sufficient for normal material. In this case the reason probably is the lack of common impurities.

A third example is the separation of sulphur. The element was trapped on a copper collector to form copper sulphide. Now it is easy enough to oxidize the material and precipitate $BaSO_4$ but many people need sulphur in a less oxidized form and the conversion of $BaSO_4$ to other compounds is not quantitative. Therefore H_2 is led over the copper sulphide and one gets H_2S which can be converted to elementary sulphur [4]. After the H_2S development stopped, we oxidized the rest of the material and precipitated $BaSO_4$. It appeared that more than one third of the sulphur was still present in the treated collector.

Apparently in all these cases the form in which the element is present either consists of two different configurations or part of the material is completely surrounded by the compound of which the collector is made.

Analysis of the Separated Material

In the separations carried out with the Amsterdam isotope separator the universal instrument to make an isotopic analysis, of course, is the mass spectrometer. We have applied some other methods as well and one of the simplest procedures is a comparison of the amounts of material which are deposited in the collector pockets. If the separation in the case of uranium is quantitative, 1 mg of U^{235} is found for every 141 mg of separated material. If the amount in the pocket for the U^{235} is 10 mg instead of 1 mg, 9 mg of the material must be U^{238} and so the concentration of the U^{235} is 10%. In this case we have supposed that the entire beam U^{235} enters the right collector pocket. It is easy enough to make a calculation for other geometries. If for instance the relative loss of the beam is equal for both isotopes and taken as x we find in the U^{235} collector $1-x+140x=139x+1$ and in the U^{238} collector $140(1-x)+x=140-139x$, so for the measured ratio of material 10 : 131 it follows that

$$\frac{139x+1}{140-139x}=\frac{10}{131} \qquad x=0.0647$$

The concentration of the U^{235} therefore is for this case $\sim 9.4\%$, i.e. somewhat lower than in the first case.

This procedure thus gives roughly the same result for both calculations and therefore is not too bad. If the original abundances of the isotopes are less different the geometry of the collector plays a more important role.

We have also applied an activation analysis for uranium which is in that case very easy. The separated material, about 1 mg, is irradiated by slow neutrons, together with a natural sample. One of the fission products of the U^{235} is Ba^{141} with a very convenient half life of 18 minutes. The Ba is separated from the material and measured. The ratio of the specific activities immediately gives the concentration of the uranium 235. This determination was done in the Amsterdam cyclotron as well as in the Kjeller reactor. Both measurements agreed to within ½ per cent with each other and with a mass spectrometer analysis made in Saclay.

Finally the very nuclear experiments for which the separated isotopes are often used may give quite a good determination of the isotopic composition. Cl made in the Amsterdam isotope separator for studying the Cl(p, γ)A reaction was compared with natural Cl and from the relative heights of some resonance lines it was possible to estimate the concentration within $\sim$5%.

This work is part of the research programme of the Stichting voor Fundamenteel Onderzoek der Materie (F.O.M.) and was made possible by financial support of the Nederlandse Organisatie voor Zuiver Wetenschappelijk Onderzoek (Z.W.O.).

References

1. O. Almén and K. O. Nielsen, Nuclear Instruments **1**, 302 (1957).
2. Zilverschoon, C. J., An Electromagnetic Isotope Separator, Thesis (Amsterdam 1954).
3. Dushman, S., Scientific Foundations of Vacuum Technique, J. Wiley & Sons (New York 1949).
4. Weaver, B., Chemical refinement procedures in the electromagnetic separation of isotopes, O.R.N.L. 1952 (1955).

Discussion

R. W. McIlroy: What method has been used to obtain elementary sulphur? We usually go through the hydrogen sulphide stage and we store sulphur as either cadmium, or some other insoluble sulphide, but we find that making elementary sulphur from sulphides is not efficient.

A. E. de Vries: We used the method described in the excellent booklet by Dr. B. Weaver. The method worked very well and we got a good material but we did not get all the sulphur out of the collector and had to apply an oxidation process for the rest of the material.

R. W. McIlroy: What was the chemical efficiency?

A. E. de Vries: I do not know the total chemical efficiency but we got $\sim$ 40% of the recovered material in the sulphide form by passing over H_2. The rest had to be oxidized to $BaSO_4$.

K. J. Hill: Dr. de Vries gave the efficiency at the collector as the weighed amount divided by the metered amount. Does he take into account the emission of secondary electrons when metering the current on the collector?

P. K. Rol: In a 180° machine the secondary emission is suppressed if the bombarded surfaces are parallel to the magnetic field lines.

R. W. McIlroy: To the question of impurities, I would like to add the following comment. In order to collect pure samples one must be careful that the beam does not strike, for example, nuts or bolts. Such components can never be obtained chemically pure, as is possible for other metal parts.

L. O. Love: In order to prevent such impurities we have done two things. Firstly we used heliarc welding instead of bolts, secondly if it is necessary to use screws we make them of graphite and purify the graphite screws in a chlorine and carbon monoxide atmosphere at a temperature of 1500–2000° C.

J. H. Freeman: There are much simpler methods in making UCl_4 than that of passing chlorine over U-metal as Dr. de Vries mentioned.

A. E. de Vries: Well, we use a very simple method for preparing samples for mass spectrometric analysis. We heat uranium oxide with C_3Cl_6. But we never used this method for source materials, only for very small amounts.

J. H. Freeman: Especially if one uses a radiated sample the C_3Cl_6 is a very efficient chlorinating agent and gives very pure UCl_4.

M. J. Higatsberger: What was the actual agreement of the two methods of analysis? I mean the method of weighing the materials in the pockets and the method using the activation analysis of slow neutrons.

A. E. de Vries: The analysis by just weighing how much material is in the pockets is generally a very crude one. But the activation analysis is very good and gives a relative accuracy of 0.5%. This is roughly the same as one may get with a mass spectrometric analysis.

M. J. Higatsberger: How accurate were the two activation analyses when the one with the cyclotron was compared to the other with the reactor? In the cyclotron you have fast neutrons. How were they slowed down?

A. E. de Vries: The slowing down in the cyclotron is done by a very large piece of graphite. The agreement between the two methods is within 0.5%.

Chapter VII

Some Health Physics Problems Connected with E. M. Separations

Protective Measures in Connection with Electromagnetic Separations of Radioactive Isotopes at the Oak Ridge National Laboratory

By

L. O. Love, W. A. Bell, J. E. Keeton, W. K. Prater, and **F. M. Scheitlin**

Oak Ridge National Laboratory, Oak Ridge, Tennessee, U.S.A.

With 5 Figures

In its inception, the electromagnetic plutonium separation facility was constructed to provide small, research quantities of the isotopes. Feed was limited and the content of isotopes other than Pu^{239} was quite low. Thus health problems were those associated with elemental plutonium, a bone seeking, alpha emitter with $2.4 \cdot 10^4$ year half-life.

With growth in program over the past years, feed levels have steadily increased in Pu^{240}, Pu^{241}, and Pu^{242}. In addition the total amount of plutonium now available for processing exceeds 500 gm and the Pu^{241} content of this material is approximately 8.5%. Commitments are outstanding to provide 10 gm of Pu^{241} having a purity within the range of 95–98% Pu^{241}. This commitment is to be fulfilled within the next operating year. Thus one must presently consider some of the aspects of isotopic plutonium when establishing operating procedures.

The Maximum Permissible Concentration (mpc) for the various plutonium isotopes is founded on recommendations of the National Committee on Radiation Protection given in National Bureau of Standards Handbook 69, issued June 5, 1959. These mpc values for the most critical organ, bone, are shown in Table 1.

Table 1. *Maximum Permissible Body Burdens and Maximum Permissible Concentrations of* Pu *in Air for Occupational Exposure*

Isotope	Critical Organ	Max. Body Burden μc	Max. Permissible Concentration in Air for 40 hour week μc/cc	μg/cc*
Pu^{239}	Bone	0.04	$2 \cdot 10^{-12}$	$3.2 \cdot 10^{-11}$
Pu^{240}	Bone	0.04	$2 \cdot 10^{-12}$	$9 \cdot 10^{-12}$
Pu^{241}	Bone	0.9	$9 \cdot 10^{-11}$	$1.7 \cdot 10^{-14}$
Pu^{242}	Bone	0.05	$2 \cdot 10^{-12}$	$5.1 \cdot 10^{-10}$

* Calculated.

With the advent of more contained Pu^{241} (>40 gm), and product samples of >95% Pu^{241}, certain additional problems arise. Sufficient alpha activity to provide adequate means for health monitoring can come only from impurities since the beta emission of 0.02 MeV from Pu^{241} lies in a very difficult energy range to permit adequate monitoring. At present we anticipate sufficient impurity, either Pu, Am^{241}, or U^{237} to provide an adequate monitoring control of product material.

The Pu^{241} included in the charge feed will be subjected to U^{237} daughter build-up, as well as to decay to Am^{241}. As pointed out by Roesch at the Geneva Conference on the Peaceful Uses of Atomic Energy, 1958, the U^{237} activity essentially reaches equilibrium in about three weeks. This time is coincident with that which we hope to achieve in one complete operating cycle. Thus Pu^{241} poses potential problems in both feed and product processing.

We should emphasize that these Pu^{241} problems have not been of sufficient magnitude to retard previous operations. We may find that they will not affect forthcoming operations too seriously. The approach to these problems will be adoption of more positive containment and provision of adequate shielding.

Isotopes of Pu^{240} and Pu^{242} exist in greater concentration in present feed than in former materials, but no new problems are anticipated as a result of these increases.

In protecting personnel from plutonium we have endeavored to contain the element totally within the operating units, units being defined as calutron vacuum tank, processing glove boxes, and wash facilities. This containment has been accomplished to the extent that radiation levels in chemical areas where feed recycle and product purification are performed do not require the use of masks or respirators. As a safety precaution, however, it is required that all personnel within the area carry respirators for use should an accidental release of activity occur. Operations in this area are shown in Fig. 1.

Fig. 1

In the calutron unit wash area, plutonium levels have required the use of respirators during the performance of any operation. In like manner all calutron unit transfers require utilization of masks or respirators. In these transfer operations plutonium is contained by a vinyl-plastic bag encasing the unit being transferred. Figs. 2, 3, 4 and 5 show some of the plastic bags required to achieve this containment. These bags are necessarily large and cumbersome to manipulate; they are difficult to join together and to separate by use of segmented septums. During certain phases of unit transfer total containment is momentarily withdrawn, creating an unprotected 1/8-in by 22-in area through which a septum is either inserted or withdrawn. The potential hazard is recognized in this operation, and the facility is

Fig. 2

presently being reviewed for the purpose of strengthening containment. It is anticipated that the calutron area will be enclosed in a room to be maintained at negative pressure as is the present wash area and the chemical area. Provided also will be a negative pressure passage way connecting the wash and calutron areas. With this arrangement, outside areas will be constantly protected by a complete double-line of defense, including initial and secondary confinement.

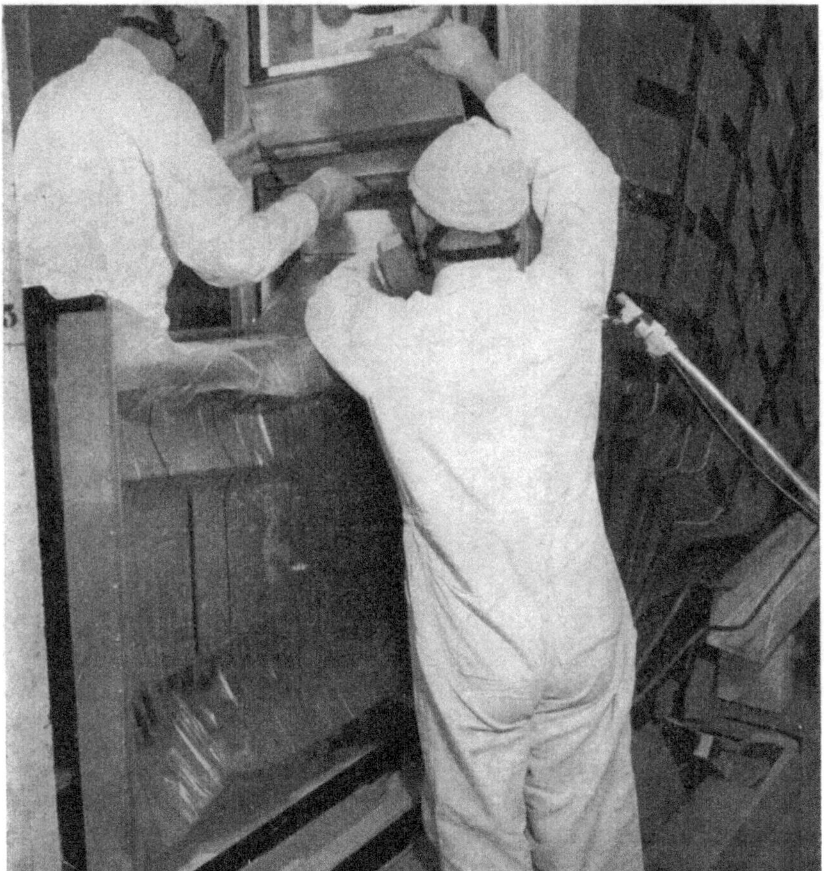

Fig. 3

Those rooms in which chemical feed recycle, product refinement, and unit wash are performed are maintained at negative pressure and both inlet and exhaust air is subjected to fiber-glass and absolute filters respectively. Appreciable activity has not been detected in these exit gases. Glove boxes and washing facilities are likewise kept at negative pressure and exhaust gases are doubly filtered through absolute filters. The glove box vacuum system is supplied with stand-by power, automatically available in case of commercial electrical power interruption.

Fig. 4

In previous operations glove box exhausts passed through a common exhaust system prior to reaching the second filter system. This exhaust was equipped with automatic fire dampers. Presently it is anticipated that the exhaust from each glove box will be subjected to a filter and a scrubber prior to entering the main filter stack; and that the main stack be provided for routine flushing to remove possible ammonium nitrate accumulation.

Fire protection is provided in the chemical areas, both by automatic activator heads and by manually manipulated valves controlling carbon dioxide extinguishers. In addition each glove box is provided with a manually operated carbon dioxide cylinder. Fire doors are provided at exits and the dampers mentioned previously are provided in the ducts to isolate connecting areas.

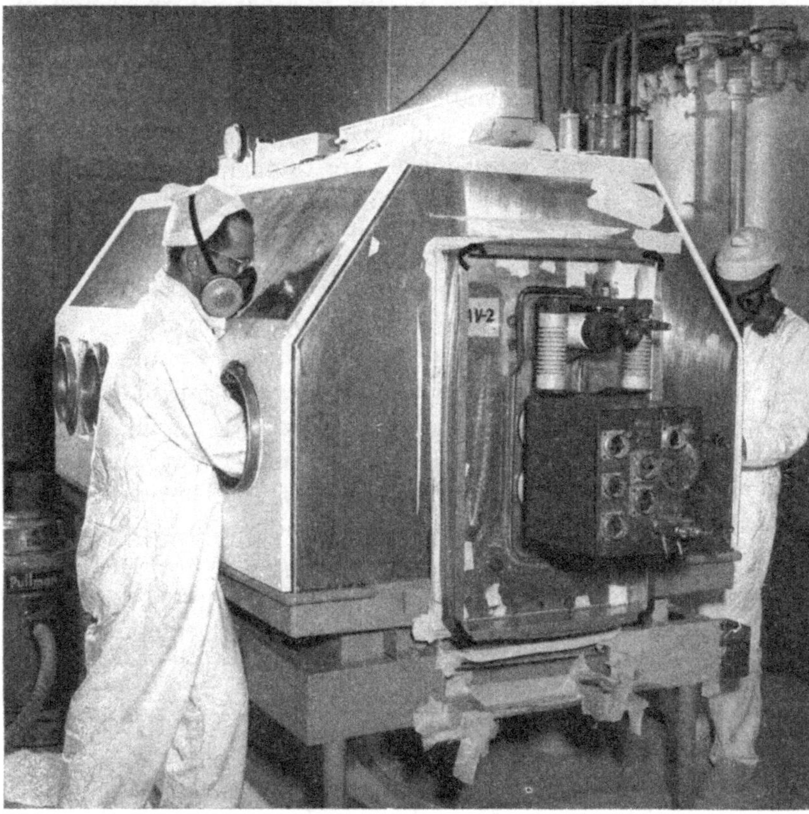

Fig. 5

Liquids containing plutonium are transferred when possible by gravity feed. Pumped liquids never exceed pressures of 35 p.s.i. Liquid wastes are packaged in plastic carboys contained in sealed steel drums. Burial is performed some 10 miles from the site of this operation.

Protective clothing is worn in all areas where plutonium contamination is reasonably possible. These consist of respirator or mask, underwear, coveralls, shoes, shoe-scuffs, etc. Personal hygiene is stressed: showers are mandatory, and smoking or eating is strictly prohibited.

Protective clothing is worn only in specified areas, and all protective clothing once worn is processed in a "contaminated laundry" facility.

Health aspects are furthered by the constant surveillance of all operations by a representative of the Health Physics Division. Through co-ordination between the groups, activities are closely monitored. Alpha surveys of personnel and equipment are performed throughout each operation with "Poppy" and "Sampson" instruments. "Cutie Pie" and "Model 2610" instruments are commonly used for beta-gamma surveys. Also utilized are film badges, dosimeter pens, and film rings. Hurst dosimeters are available for neutron surveys.

Each area of operation is provided with one or more suitably located air samplers. From these stations two air samples are removed daily. A chart-type alpha monitor is also provided in the chemistry area. This instrument is capable of audible warning should a sudden rise in activity occur. However, there exists a possible time lag between occurrence of activity and alarm depending upon the read-out frequency rate — generally in excess of 10 minute intervals. Thus the development of a low-level, continuous alpha air monitor is to be encouraged. Air monitoring on a less rigorous but routine schedule is performed on all main exhaust ducts. Liquid wastes are also subjected to sampling prior to disposal. Floor areas, glove boxes, transfer equipment, etc., are routinely subjected to smear counts to determine the degree of surface contamination.

Personnel monitoring is performed continually during all transfer and unit wash operations. Hands are periodically checked for activity which might have passed through the rubber glove and the inner protective cotton glove. Shoe-scuffs are likewise checked, as are floor areas most subject to become contaminated. When a transfer operation is completed all protective clothing is surveyed for alpha activity before it is removed. Occasionally a defective glove results in cotton glove contamination, but rarely results in contamination to the hands themselves. Urine samples are required of personnel working in plutonium areas.

Activity levels normally encountered in the chemical, the wash, and the calutron track area are presented in Tables 2, 3, and 4 respectively.

Table 2. *Alpha Activity Levels Present in Chemical Processing Areas*

Area	Date Sampled	D/M/M³
Main Filter Exhaust	10–2–59	0.4
	10–13–59	0.0
	10–16–59	0.7
Chemistry Laboratory	10–2–59	0.3
	10–15–59	0.9
	10–20–59	0.5

Table 3. *Alpha Activity Released in Wash Area During Operations*[1]

Typical Operation	Minutes	D/M/M³
Unit Transfers	105	14.1
	96	3.1
	170	27.2
Equipment Repairs	48	218.0
	85	4.9
Unit Wash	412	7.0

Table 4. *Alpha Activity Released During Calutron Unit Transfers*

Typical Transfers of Units	Distance of Sampler from Operating Area					
	1 foot		3 feet		50 feet	
	Min.	D/M/M³	Min.	D/M/M³	Min.	D/M/M³
Liner	89	66.5	89	10.0	89	1.3
	75	392.0	60	27.2	60	6.3
	210	88.0	210	0.3	210	0.2
Source or Receiver	198	5.1	198	60.9	198	1.4
	99	0.2	93	0.0	92	0.0
	85	39.6	105	14.7	85	2.1

The Maximum Permissible Concentration of air-borne plutonium is $2 \cdot 10^{-12}$ nc/cc per 40 hour week. This corresponds to 4.4 disintegrations per minute per cubic meter of air. The detection of one mpc of plutonium in sampling atmospheric dust deposited on a filter disk is complicated by the activity resulting from naturally occurring radon

[1] Samples taken approximately 2 feet from operation.

and thoron. Furthermore, almost any operation associated with calutron unit transfer is completed long before sampling results are available. Because of this fact each operation is approached as if mpc were to be exceeded.

Liquid spills involving small volumes of solution have occurred. Prompt decontamination measures have kept the plutonium confined to the area involved in the spill. Adequate scrubbing and painting has generally sufficed to provide decontamination; however, complete removal of activity sometimes may require replacement of approximately two inches of concrete.

In summary, the approach attempted is total containment of plutonium, allowing personnel maximum freedom from discomfort in work areas. Alpha activity surveys indicate that total containment in certain operations and areas is not achieved, and that mpc for 40 hour week is exceeded. To provide protection in these areas, personnel are required to utilize masks, monitoring, etc., to prevent bodily intake of air-borne plutonium. The method used in effecting this degree of containment is cumbersome, yet it is sufficiently effective to permit the accomplishment of isotopic separations without incident.

Discussion

J. Kistemaker: What is the vapour pressure of Pu and PuO_2 and how dangerous are these materials in the open air? The health limit given by Dr. Love for Pu^{241} seems to be about 10^{-13} atmospheres.

J. H. Freeman: In my opinion the hazard is a dust hazard arising mainly from loose powder on the source and collector units.

P. Weinzierl: How is the low alpha activity measured corresponding to the permissible level during the presence of a much higher natural alpha-activity level in the aerosol?

R. W. McIlroy: It is essentially a dust analysis. The system used is to draw some m^3 of air through a filter paper over a period of 8 hours and then ash the paper and count it.

P. Weinzierl: If you wait long enough so that the natural activity of radium products in the air is decayed then your alpha-activity will practically be Pu, but if you count immediately you will have much more natural alpha-activities than from Pu.

P. K. Rol: This natural activity can be cancelled out by measuring also the ratio between alpha- and beta-activity. When Pu is added the alpha-activity increases and also the ratio between alpha- and beta-activity changes.

L. O. LOVE: We did this the same way. We have often wondered what would happen if one just pulls the Pu-source out and leaves it. It picks up moisture like the other rare earths and it sticks. Dr. FREEMAN mentioned that it was fluffy but it is only fluffy when it is dry.

R. W. McILROY: We have taken air samples inside the Hermes dry box at the time during the changing of the source unit. These air samples show 100 x of the permissible concentration.

J. H. FREEMAN: We have seen light powdery deposits on Pu-source units. All of our experiences on Pu chemistry indicate that recoil mechanisms do not play a serious part in activity spreads.

H. J. LANGMANN: Do you have to be concerned with criticality conditions when separating Pu?

L. O. LOVE: Yes.

Precautions with α-Active Isotope Separators at Harwell

By

R. W. McIlroy

Chemistry Division A.E.R.E., Harwell

With 7 Figures

There are three main alternatives in setting up an α-active separator. At one extreme the whole machine can be boxed up and serviced

Fig. 1

by men in fully protective clothes, whilst at the other an open machine can be used with foolproof bagging arrangements for changing units at the operating face. An intermediate system can be envisaged in which the machine is boxed but has its main units close to walls containing glove ports, or even in individual dry-boxes.

At Harwell the former alternative was chosen for Hermes. This machine, which has been described in detail before (1–5), is a four-foot radius 90° separator intended specifically for plutonium and possibly higher transuranic separations. Although there is some external radiation, particularly with highly irradiated sources, the main hazard in work with plutonium is one of inhalation or ingestion. The weak γ associated with Americium 241 has not so far caused any trouble in Hermes itself but it does affect the handling of old source material (1–2 years old) in bulk. The machine is enclosed in a sealed room of 500 sq. ft. joined on one side to the pressurised suit area of the Radioactive Chemical Laboratory (a service already available when Hermes was built), and having the control room on the other side and separated from it by large double perspex windows giving

Fig. 2

a good side view of the machine from the control desk (Figs. 1 and 2). All operations on the machine, which include maintenance of services and changing of source and collector units (but not the assembly of units) are carried out by men in pressure suits (Fig. 3) controlled from the Chemical Laboratory Service Centre but in telephonic communication with the Hermes control room. The sealed room is kept at 0.2″ water pressure below the outside pressure and access is only possible through a controlled changing and washing room.

Fig. 3

Services

Low tension electrical connections are made by sealed pyrotenax leads (copper cores in a copper sheath insulated with compressed magnesium oxide; the sheath can be brazed to mounting plates). H.T. leads pass through sealed porcelain insulators.

Pipe entries are kept to a minimum but three gas leads are provided for feeds to the machine, each fitted with a non-return valve and filter in case of accidental reversal of flow, and one evacuation line for the liquid nitrogen vacuum jacket. The pump for the line is outside the dry-box but the exhaust is directed back inside.

Deionised cooling water flows in closed circuit and an ion exchanger is provided outside the box to deal with any dissolved solids which may accumulate in the water either from normal corrosion or by active contamination. No hazard has developed from pyrophoric plutonium, but automatic carbon dioxide fire extinguishers are fitted.

Control Room

There are no openings between the machine room and the control room and this can therefore be regarded as relatively "clean", with a minimum of restriction to access from outside. This has considerable working advantages as the extensive electrical apparatus associated with Hermes requires continual maintenance and all work is simplified if the routine restrictions for work in active areas can be eased.

Servicing

Within the sealed area there are a limited number of machine tools available and a fairly wide range of maintenance work is possible without the excessive decontamination of parts which would be required to bring them into an open workshop. Under normal operating conditions collector units are changed bodily after three to five runs and source units are changed as necessary, generally when the charge runs out. This is timed to be rather sooner than when the filament would fail, and is usually after about two days operation. The whole assembly is cleaned after several months' running. Waste material, wash liquors and used collectors are all passed out of the pressure suit area, suitably bagged in polythene, and are worked up in conventional glove-boxes in the chemical laboratory. New collectors are assembled in clean areas (most parts are used only once) but final attachment to the collector face plate must be done in the sealed room. This operation takes about one hour. Source units are used many times, however (only the filament and graphite parts of the electrodes being renewable at each reloading), and as the operation of stripping, cleaning, and reassembly is rather dirty and slow (taking about 8 hours) a special glove-box is available between the pressure suit area and the radiochemical laboratory (Fig. 1, top right hand corner). The source unit can be placed in this from the active side and serviced by normal glove-box techniques from the laboratory side. This reduces the work done in pressure suits to the actual fitting to the machine, which takes about one hour. All active units are bagged as soon as they are removed from the machine and since the dirtiest operation (cleaning the source unit) is confined to the small box, the general level of activity even in the sealed room can be kept comparatively low (5 mpc normally) but it may rise sharply to as much as 50,000 mpc when changing source units. During the major clean up at several month intervals all operations are done by wet processes and so little spread of activity results. The advantages of this system are:

i) Elaborate bagging of large and clumsy units is unnecessary.

ii) The extreme care necessary to eliminate the dust hazard completely is reduced to keeping a small area relatively clean without the necessity of absolute cleanliness.

iii) The control room and associated electrical gear are almost free from restriction of movement of men and materials.

iv) Pressure suit working is kept to a minimum by the use of adjacent glove-boxes.

The major disadvantage is the excessive time required to carry out minor adjustments — it takes about 20 minutes for a man to enter or leave the pressure suit area. A great deal of automatic operation is therefore necessary — all vanes, probes, gas valves and common adjustments are motorised and though in general these work well more automation always means more routine maintenance.

A secondary disadvantage is the lack of a direct view of the beam at any point. In the calutron, windows enable the source and collector units to be examined when necessary by the operators and on balance this is a useful facility. We have used a closed circuit television link but this has not proved as satisfactory as direct viewing and must be supplemented by a relatively large number of devices for locating the beam and measuring its intensity distribution. These disadvantages are mainly associated with the trial periods which follow major shut downs and modifications. When running smoothly under routine conditions they are less severely felt and at these times the convenience of the sealed room method shows itself in the rapid turn round of source units. From the end of one run to the beginning of pumping down for the next can be as short as three hours (a second source unit being made ready in advance). A quick turn round is essential in Hermes since the charge is small (ca. 10 grammes).

Chemical Services

Source material is prepared and recovered, and isotopes are handled in a separate suite of glove-boxes by normal methods. A special problem arises, however, with nearly pure Pu^{241}. This is β-active with an energy of only 0.02 MeV and a half-life of 13 years. The daughter is Am^{241}, which is α-active with a half-life of 470 years. It is felt that if Pu^{241} is deposited in bone the daughter will be fixed by biological processes in the same place and the hazard from Pu^{241} is therefore that from Am^{241} assumed to be following the plutonium metabolic path. This sets a lower limit of acceptable contamination than would arise from Pu^{241} alone. Monitoring, however, must be carried out on the plutonium itself and this poses a difficult problem since the β-emission has a range of less than 1 mg/sq. cm and no windowed counters are available for this energy, and therefore no direct surface monitors. Reliance must be placed on extra care in handling coupled with the usual supporting operations of urine analysis, air sampling and surface smears, all of which will give evidence of accidents only some time after they have happened. With 95% Pu^{241} there is just enough α-emission from other isotopes to get a reading of 1 mpl of Pu^{241} on a standard α hand monitor set to operate for seven times the normal time. In order to prevent the effect being masked by other α-contamination the hands are counted both on entering and leaving the laboratory and all other α-activity is excluded from the room. A further modification to normal procedure is to monitor the hands before washing. This will give immediate warning of an accident which pre-washing might have obscured. All work on this isotope is confined to the special laboratory and to a minimum of people. Entry to the laboratory is restricted and entails a separate change of clothing apart from the change made on entering the building. Normal working rules are observed with emphasis on care, and in addition glove-boxes are always fitted with double gloves, of which the inner one is changed every month, compared with the normal period of several months. Smears are taken periodically and air samples are taken whenever operators are in the room. We have so far been successful in our precautions and there has been no unsuspected release of activity from the glove-boxes at all.

Open Operations

As an example of what can be done with simple equipment on an open machine there follow the precautions used in a separation of enriched U^{235} in the calutron. To obtain very pure U^{235} with a minimum of effort we used a highly enriched source material and

Fig. 4

Fig. 5

although 235 is not particularly hazardous in normal use the dusty nature of calutron operations and the normally unhygienic design of large plant in non-active areas made precautions necessary.

The area around the working face of the calutron was entirely enclosed in a light, temporary structure of P.V.C. sheet stretched on Dexion frames (Figs. 4 and 5) and access was restricted through a changing room where light pressure suits could be donned (Fig. 6).

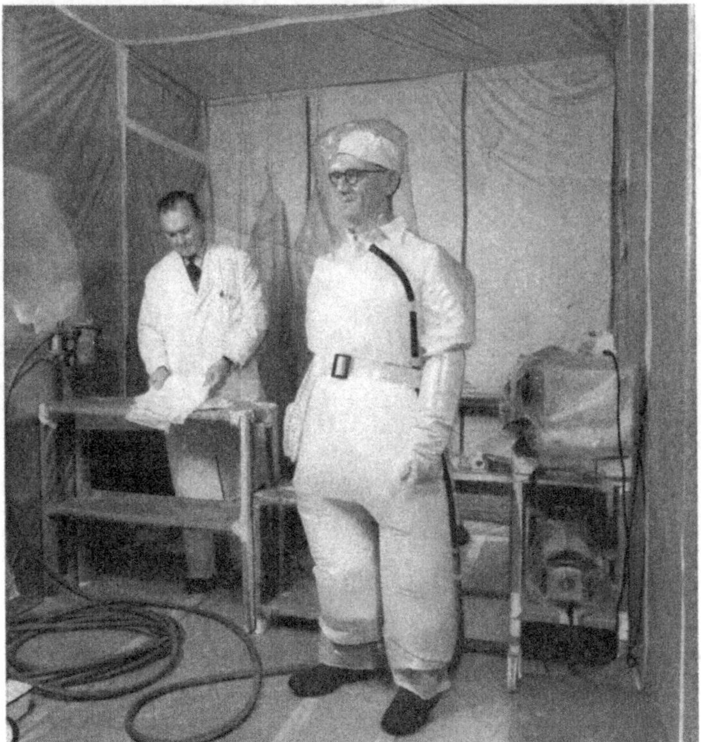

Fig. 6

These were obligatory for operations involving removal of a unit, and for other work an ordinary clothing change was sufficient. As much of the permanent equipment on the machine as possible was protected by tape and P.V.C., and the floors were covered. The partition confined dust to a small area and this was kept clean and continually monitored as in normal active working. Units were bagged on removal and serviced in a smaller enclosed box (Fig. 7). As a result of careful working no external parts of the machine had contamination greater than 4 mpl at the end of the run, lasting several weeks, and after a general clean up of the area it was possible to remove the partitioning within 10 days. No contamination had to be dealt with outside the screened area and although during source changes the air within the

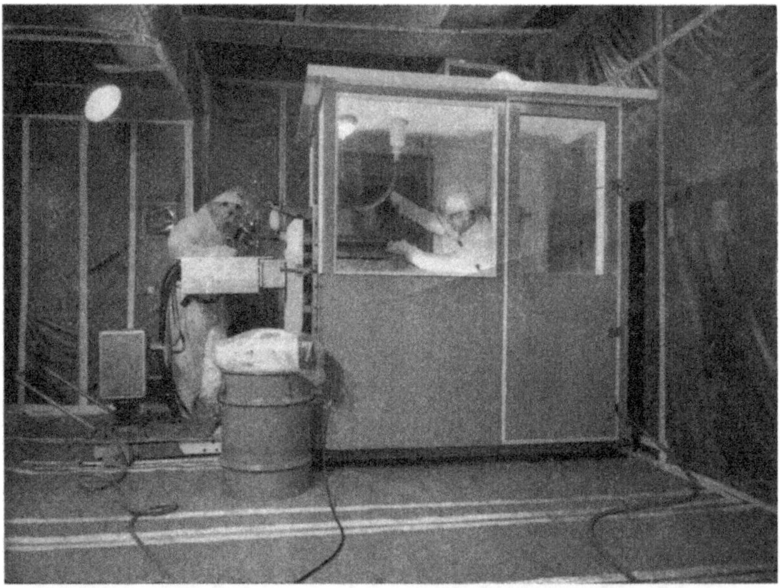

Fig. 7

screen contained up to 40 mpc of uranium the general level outside never rose above 1 mpc. This degree of care, which was close to the maximum possible without elaborate bagging procedure, is quite feasible for rich U^{235} but would be insufficient for anything more active, or with greater specific hazard.

The Third Alternative

Between the two extremes of bagged units on an open machine and a completely boxed machine, the third possibility, consisting of placing the machine in a sealed room (which is necessary at times of dismantling for cleaning) but bringing the source and collector units to glove-boxes in the wall of the room, has the disadvantage of making the operating room a potentially active area, unless separated from the machine by yet another partition. It would also add to the mechanical difficulties, since the necessity of removing units bodily from the machine cannot be entirely avoided.

Our experience suggests that the Hermes arrangement has advantages over any alternative system and that the combination of relatively easy access to the machine with the use of separate glove-box working for long and tedious operations has much to recommend it. This, however, is only for α-active materials. It does not necessarily remain true for γ-active isotopes with which it may be impossible to take full advantage of the large working space round the machine since access may be restricted for other reasons.

References

1. FREEMAN, J. H., and M. L. SMITH, Glove-boxes and Shielded Cells, p. 306 (Butterworth 1957).
2. FREEMAN, J. H., K. J. HILL, and M. L. SMITH, E.M. Separation of Plutonium Isotopes, Operational Experience in 1957, A.E.R.E. Report No. R-3043.
3. DAWTON, R. H., Proceedings of the Harwell Symposium on Electromagnetically Enriched Isotopes and Mass Spectrometry, p. 208 (Butterworth 1956).
4. SMITH, M. L., and K. J. HILL, Proc. International Symposium on Isotope Separation, p. 581 (North-Holland 1958).
5. Electromagnetic Separators, p. 150 (North-Holland 1958).

Extra Remarks

1. The active material which collects in Hermes tank and which has to be recovered periodically for criticality reasons is distributed roughly in the proportions:

Source unit	40%
Source tank	40%
Flight tank	10%
Remainder	10%

Very little finds its way into the diffusion pumps. Of the 10% in the flight tank almost all is firmly embedded and can only be removed by processes which remove the metal surface.

2. Recovery of isotopes from collectors by scraping requires great care since it is easy to contaminate the sample with iron from the scraping tool. If graphite of sufficient purity can be obtained (50 ppm of iron or less) it is better to burn collectors without scraping.

Chapter VIII

Applications of E. M. Separated Radioactive Isotopes

Applications of Electromagnetically Separated Radioactive Isotopes at the Nobel Institute of Physics

Part I

By

Torsten Alväger
Nobel Institute of Physics, Stockholm

With 2 Figures

The radioactive isotopes prepared with the E.M. isotope separator at the Nobel Institute of Physics [1] have been used in a large variety of investigations. Table 1 gives an idea of the problems which are of

Table 1

Type of problem	Advantage of use of isotope separation	Isotopes of element	Ref.
A Short half-lives (ms-ns region)	Mass number determination Decreased number of chance coinc. Decreased number of prompt coinc.	Bi, Pb, Tl, Cs, Xe, Br, Kr	2, 3, 4
B Very weak transitions such as "double quantum emission"	Possibilities of avoiding impurities of different elements which are difficult to separate chemically	Xe, Sr, Te	5
C Heavy ion reactions (C^{12}, Ne^{20})	Mass number determination	Bi, At, Po	13, 14
D Nuclear spectroscopy of fission products	Mass number determination Pure isotope sources	Sn, Sb, Te, I, Xe	6, 7
E Low energy β-spectroscopy	Thin sources	Bi, Pb, I	8

current interest and those isotopes which have been separated. They fall mainly into five groups:

A. Measurements of short half-lives by use of delayed coincidences,
B. Search for very weak transitions, such as double quantum emission,
C. Mass determination of products obtained in heavy-ion reactions,
D. Nuclear spectroscopy of uranium fission products and
E. Low energy β-spectroscopy.

The elements separated are mainly grouped around the double filled shells $Z = 82$, $N = 126$ and $Z = 50$, $N = 82$. Not many separations have been made of the rare earth elements for example.

We shall now discuss mainly some details concerning the groups A, B, and C, while D and E will be considered in part II.

A. Measurements of Short Half-Lives of Excited States

In these investigations gamma-scintillation-spectroscopy has been used as this is the simplest and most efficient method for gamma-detection. The time resolution 2τ obtainable is often not larger than a few ns for low energy gamma-rays. Most transitions will therefore appear as "prompt", i.e. they have a half-life shorter than can be detected.

As a certain isotope of interest cannot usually be produced isolated from other isotopes, the advantages of isotope separated samples in this case are manifold and often separation is essential. Firstly, one gets a mass number estimation of the isotope to be studied. Further, as the energy resolution of a scintillation spectrometer is relatively poor, one has often a very complex spectrum of unresolved lines if no isotope separation is performed. This complexity together with the ever present Compton background, is the cause that, even if the spectrometer is set on two photolines which should give a delayed coincidence curve, possible prompt coincidences from unresolved lines and Compton background from isotopes other than the one of interest, can completely mask the effect to be studied.

The third advantage of isotope separated samples in experiments of this type is that the ratio between true (C_r) and chance (C_c) coincidences can be increased considerably. As $C_r / C_c = (2\tau \cdot S)^{-1}$, where S is the part of the source strength which corresponds to the focused energy interval used, one can, by E.M. separation, greatly decrease disturbing γ-rays and therefore increase the ratio C_r/C_c. This is of special importance for investigations where one measures coincidences between X-rays and a photoline, which is a convenient method when searching for new isomeric states.

Following the procedure discussed above, half-lives of several isomeric states have been measured in isotopes of Bi, Pb, Tl, Cs, Xe, Te [2, 3, 4].

Fig. 1 gives a typical example. The half-life of the 4 ms level in Pb^{205} was measured by taking coincidences between X-rays and the γ-ray from the 4 ms level [4]. It should be pointed out that in this case it was not absolutely necessary to use a separated sample but the separation did give big advantages. Often, however, an isotope separation has been necessary in the measurements discussed above.

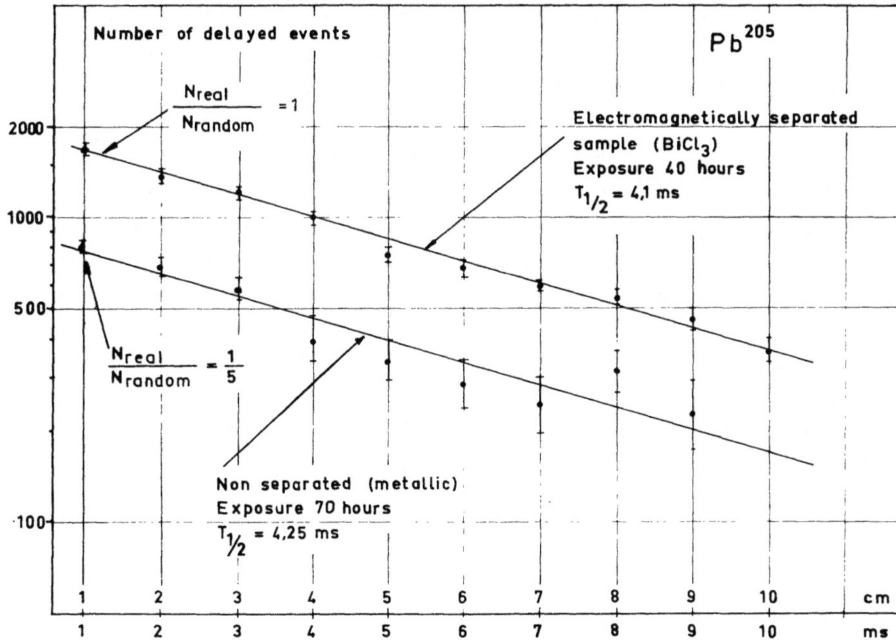

Fig. 1. Decay curves of the $i_{13/2}$ state in Pb^{205} showing the advantage of using electromagnetically separated samples in delayed coincidence measurements (from [4])

B. Weak Transitions

When studying very weak effects one must naturally be careful to eliminate the possibility that the observed effect can be due to some kind of contamination of the sample. As it is often difficult to achieve high chemical purity of an element, there are often many possibilities of contamination. In such a case it can therefore sometimes be suitable to use an isotope separation, thus eliminating chemically similar elements with often quite different Z value.

A typical example of weak effects which we have studied recently and which has not been observed earlier is the double quantum emission. This is a process where an excited level decays by two gamma-rays of energies E_1 and E_2, respectively, instead of one single gamma-ray of energy $E=E_1+E_2$. The energy spectrum of E_1 and E_2 varies continuously. As the ratio between the transition probabilities for double to single gamma-emission can be of the order of 10^{-5}, it is obvious that small contaminations of isotopes even of elements other than the one of interest can give false results. The isotopes which are studied at present are Xe^{131m} [5] and Rb^{85}. By an isotope separation of these, one is quite sure that one really has no other mass number present which could contribute disturbing effects.

C. Heavy-Ion Reactions

The heavy-ion reactions investigated recently, with special emphasis on mass number determination of the reaction products obtained, are of the type $A(C^{12},xn)B$ where A stands for Ir, Pt, or Au while B is Bi, Po, or At. Also Ne^{20} reactions have been investigated to a certain extent.

The problem of mass estimation of products from heavy-ion reactions is not very different from similar investigations on, for example, spallation and fission reactions. The yields of heavy-ion reactions are, however, often very low, mainly due to both small cross-sections and low intensity of the bombarding ion beam. Many of the isotopes studied are produced to a total of only a few thousand atoms.

This fact placed high demands on the isotope separation. A total efficiency of at least 10% is usually needed in a short time separation. Further the half-lives of the isotopes produced are often short, often less than 15 min. This means that it is necessary to use a fast separation method, such as, for example, that discussed earlier in which two ovens are used in the ion source [9].

The heavy-ion reactions were performed in the large cyclotron at the Nobel Institute of Physics [10]. The targets had a thickness of about 20 mg/cm². For collecting the radioactive products the recoil technique was used. Catcher foils of tygon (0.7–0.8 mg/cm²) were placed close behind the target foil. After the irradiation the catcher foils were placed directly into a Pt-oven which was then used in the isotope separator. Pb was used as carrier in the separator. The time necessary for the whole separation procedure was only a few minutes. The different isotopes separated were detected by α-spectroscopic measurements. Activities with half-lives as short as about 5 minutes were studied in this way.

Fig. 2 shows an example of an α-spectrum obtained from a Po sample and measured with a multichannel analyzer. The curve A is taken of a non-isotope-separated sample [11]. The different energies measured are ascribed to a certain mass number only from systematics and are therefore rather uncertain. The curve B gives the α-energy

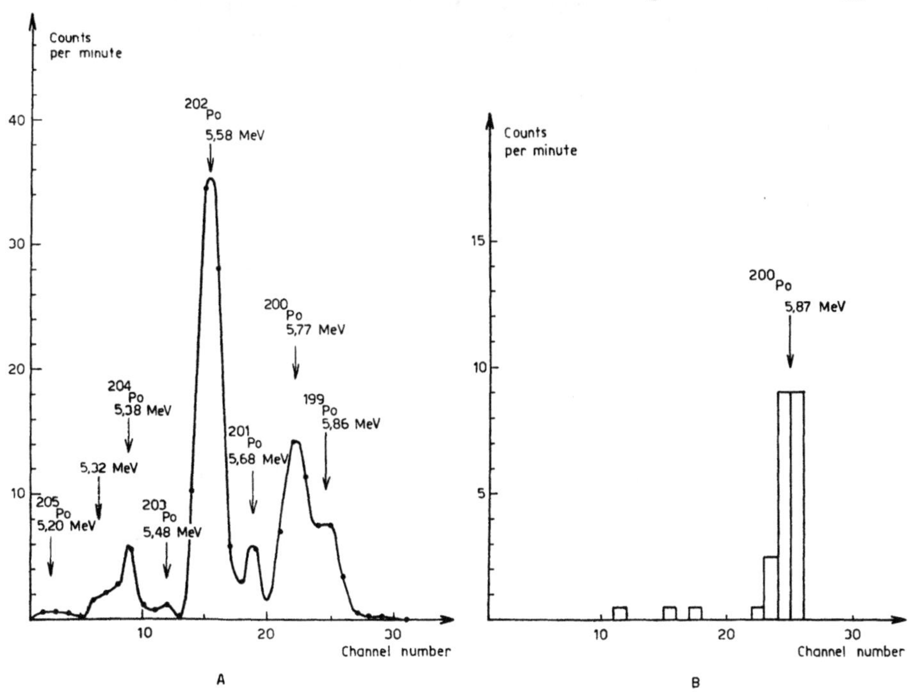

Fig. 2. α-spectrum from Po isotopes obtained in the reaction $Pt(C^{12}, xn)Po$. Curve A is from an non separated sample while B is from a separated Po^{200} sample

due to one of the mass-separated Po-isotopes, Po^{200}. This energy was earlier thought to be due to Po^{199}, while the α-group assigned to Po^{199} disappeared in the isotope-separated samples. This shows the large hazards in mass estimation without the use of isotope-separated samples.

The possibility of mass number estimation of activities with such short half-lives and small creation cross-section as discussed above is mainly due to the fast separation procedure with the two-oven construction of the ion source. The experiments are also examples of how short-lived isotopes can be investigated when one has the accelerator and separator at different places. To study even shorter lived activities than those given above it seems necessary to produce the isotopes directly in the ion source of the separator (cf. e.g. [12]).

References

1. Alväger, T., and J. Uhler, Ark. Fys. **13**, 145 (1958).
2. Alväger, T., B. Johansson, and W. Zuk, Ark. Fys. **14**, 373 (1958).
3. Johansson, B., T. Alväger, and W. Zuk, Ark. Fys. **14**, 439 (1958).
4. Bergström, I., and al., Nuclear Instruments to be published.
5. Alväger, T., and H. Ryde, Ark. Fys. **17**, 535 (1960).
6. Alväger, T., J. Uhler, and O. Melin, Ark. Fys. vol. 16, no. 42.
7. Alväger, T., Ark. Fys. **17**, 521 (1960).
8. Stockendal, R., Ark. Fys. **17**, 552 (1960).
9. Uhler, J., and T. Alväger, Ark. Fys. **14**, 473 (1958).
10. Atterling, H., Ark. Fys. **15**, 531 (1959).
11. Atterling, H., and W. Forsling, Ark. Fys. **15**, 81 (1958).
12. Koch, J., Production and Utilization of E. M. Enriched Isotopes (Harwell 1955).
13. Forsling, W., T. Alväger, L. W. Holm, O. Melin, J. Uhler, B. Åström, to be publ. in Ark. f. Fysik.
14. Forsling, W. and T. Alväger, to be publ. in Ark. f. Fysik.

Discussion

J. H. Freeman: I would like to know whether Dr. Alväger's work with C^{12} reactions can be extended to identify the higher transuranic elements?

T. Alväger: There seem to be no definite difficulties in using the technique used in the mass assignment of isotopes produced by C^{12} reaction in the transuranium region.

H. J. Langmann: Was the quoted figure for the double quantum decay of Xe^{131} calculated or measured?

T. Alväger: The ratio of 10^{-5} given for a double quantum emission was a rough, theoretical value. The measured value for Xe^{131} was of the order of 10^{-4} as estimated from X-ray — X-ray coincidences measurements.

Applications of Electromagnetically Separated Radioactive Isotopes at the Nobel Institute of Physics

Part II

By

Jan Uhler

Nobel Institute of Physics, Stockholm

With 4 Figures

The principle aim of our review of work with separated samples at the Nobel Institute of Physics is to show the general applicability of separation of radioactive material as a method in nuclear spectroscopy.

Two examples will be given here. They are intended to show that the existence of the separator can encourage research work on problems otherwise regarded as difficult. Compare Part I with the same title.

Low Energy β-Spectroscopy

Firstly, reference will be made to a specific and well-defined problem constituting a small part of the systematic investigations of odd lead isotopes that have been undertaken for many years at the institute.

According to nuclear shell model predictions, gamma transitions of the type $i_{13/2} - f_{5/2}$ are expected to take place in odd mass nuclides in the vicinity of the double-magic nucleus Pb^{208}. Such transitions have been found and measured for lead isotopes, forming a systematic trend with respect to energy and metastable half-life. For Pb^{205}, daughter of Bi^{205}, however, no such transition fitting the systematics was found and this was explained by the existence of some intermediate level.

However, the existence of an isomeric transition having too short a half-life and the wrong multipolarity was quite confusing. One possible explanation was the existence of a low energy, highly converted transition. To detect such a transition there is a need for precision energy measurements of conversion electrons in the energy region 2–50 keV. Such measurements require a very thin sample without any other disturbing high activity, which in this case would mainly be Bi^{206}.

The irradiation was made in the 255 cm cyclotron of the institute on radiogenic lead (88 % Pb^{206}, 9 % Pb^{207}, and 3 % Pb^{208}). The bismuth activity was chemically separated and then electromagnetically separated as the chloride. The activities were collected into thin aluminium foil (about 1.9 mg/cm²). Fig. 1 shows an autoradiogram. This is the original negative, laid on the foil with the separated activities. This is one of the methods we use for determining the exact position of activities. The lines here appear to be more diffuse than their actual shape, but the determination of the mid-point is very accurate. In fact, this separation consists of two different runs with the same collector and, as seen on the picture, the focusing of the mass lines was well reproduced from one run to another. The β-source, a narrow strip cut from the collector foil, had dimensions which made them very convenient for measurements in a double-focusing β-spectrometer. In measure-

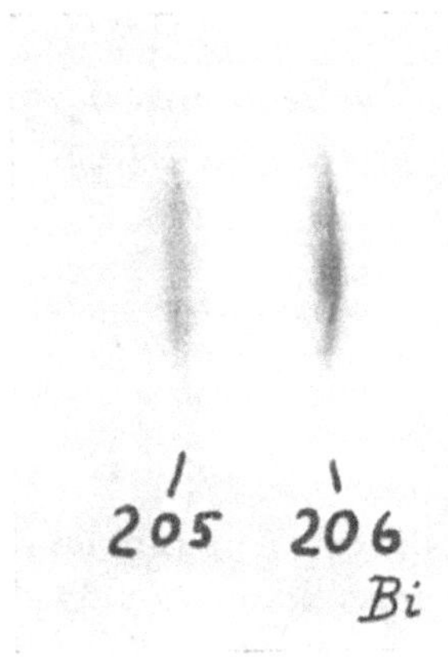

Fig. 1

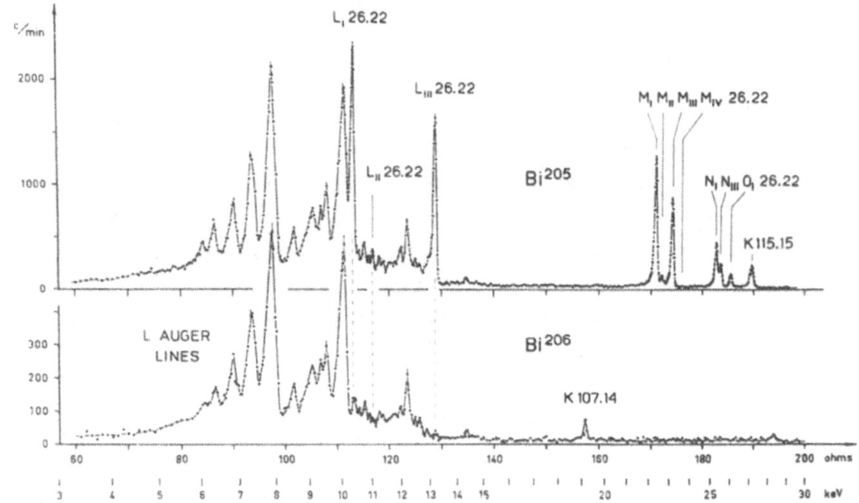

Fig. 2

ments of very low energy electrons one uses very thin G.M.-counter windows with a cut-off energy of about 2 keV. To utilize this, it is very important that the sample is as thin as possible, since any deposit would give a broadening of the lines and spoil the chances of classifying the lines, especially in the Auger region.

Fig. 2 gives the low energy part of the spectrum in Bi^{205} and Bi^{206}. The simplest way of classifying the Auger lines is to compare these spectra and thus the strong L- and M-converted 26.22 keV lines were found and classified as due to a magnetic quadrupole transition. The existence of this transition explains completely the observed isomeric half-life. Although this sample is less than about 100 atomic layers thick this gives a broadening of the low energy lines by a factor of 2.

One thus sees in this investigation the way in which results can be obtained by application of the separator to a suitable problem, in this case a very intense transition. However, in the search for weak lines in many cases one cannot permit the loss of activity that follows isotope separation [1].

Nuclear Spectroscopy of Uranium Fission Products

There still exist regions of the periodic system where only very few nuclear investigations have been performed. One of these domains contains the neutron-rich tin, antimony and telurium isotopes grouped around the double-magic nucleus Sn^{132}. The main reason for this lack of information is that the best way of producing them is by nuclear fission. Another reason is that many of them have shown to have half-lives in the isomeric or ground state of about the same magnitude.

Table 1 shows the region and what kind of information has been published up to date. In many cases the degree of certainty of each isotopic assignment is indicated by the letter D, which means "element certainty and mass number not well established."

Our first problem was thus to give mass determinations of observed half-lives. As far as we can see now, most assignments are correct but precise measurements of half-lives have not been performed on all isotopes found.

At the same time spectroscopy has been done using scintillation counters. Thus, measurements of γ-rays with multichannel analyzers and coincidence measurements between intense γ-transitions have been carried out.

After an electromagnetic separation it is always important to detect the activity and estimate the exact position of activities on the collector as well as the relative yield of different isotopes.

Table 1

	Sn	Sb	Te	I	Xe	Cs	
125	A 9.4 d β^-,γ A 9.5 m β^-,γ	A 2.0 y β^-,γ	A 58 d γ stable				125
126	B 50 m	D 30 d β^- D 19 m γ B 9 h β^-,γ	stable				126
127	B 2.1 h	(?) D 6.2 d γ A 90 h β^-,γ	A 105 d γ A 9.4 h β^-,γ	stable			127
128	D 57 m	D 9.6 h γ D 10.3 m β^-,γ	stable				128
129		A 4.2 h β^-,γ	A 33.5 d β^-,γ A 74 m β^-,γ	A $1.7 \cdot 10^7 y$			129
130	D 2.6 m	D 40 m D 9.2 m	stable				130
131	B 3.4 m	A 23.1 m	A 30 h β^-,γ A 24.8 m β^-,γ	A 8 d β^-,γ	A 12.0 d γ stable		131
132	B 2.2 m	B 2.1 m	B 77.7 h β^-,γ	B 2.26 h β^-,γ	stable		132
133		B 4.4 m	A 53 m γ A 2 m β^-,γ	A 20.8 h β^-,γ	A 2.3 d γ A 5.27 d β^-,γ	stable	133
134		D 50 s	B 44 m	B 52.5 m β^-,γ	stable		134
135			[A] < 2 m	A 6.7 β^-,γ	A 15.6 m γ A 9.13 h β^-,γ	A $3 \cdot 10^6 y$	135
136			E ~ 1 m	D 86 s β^-,γ	stable		136
	Sn	Sb	Te	I	Xe	Cs	

The autoradiogram has already been mentioned. Another way of measuring the distribution across the target is by cutting part of the target foil into small strips and measuring the activity on each strip.

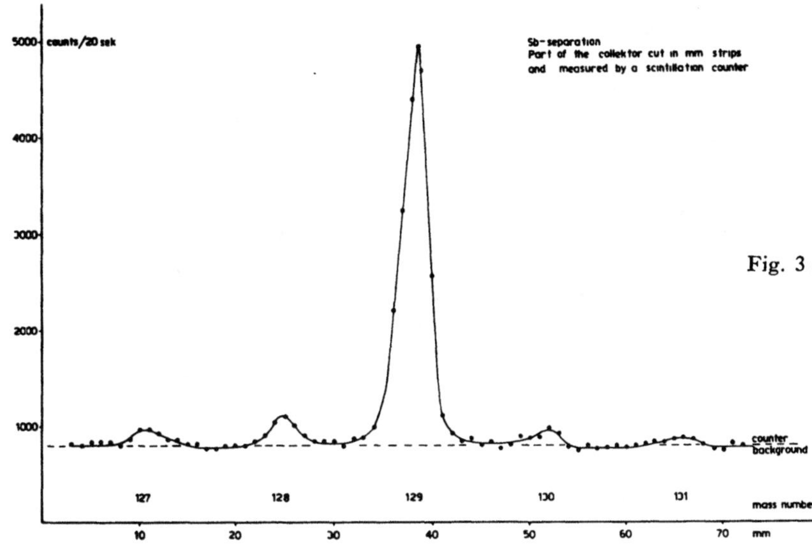

Fig. 3 a

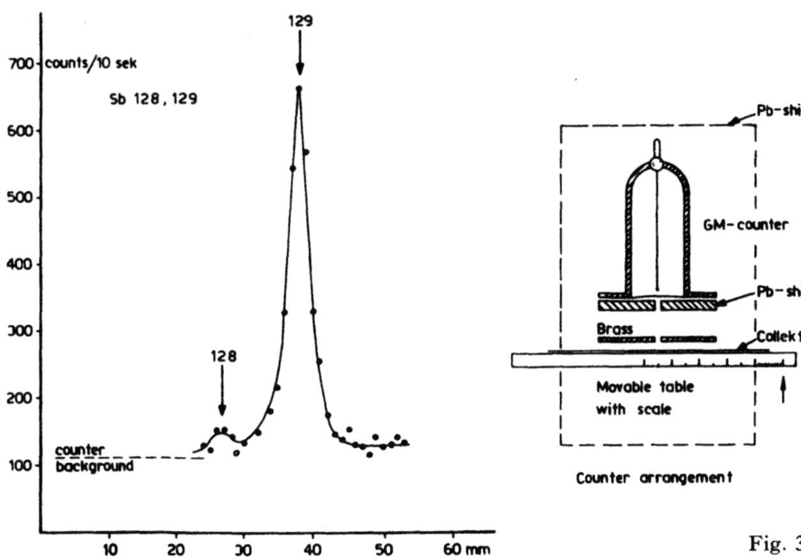

Fig. 3 b

Here, in Fig. 3 such a picture for a Sb separation is shown. The dominance of mass number 129 emphasizes the importance of isotope

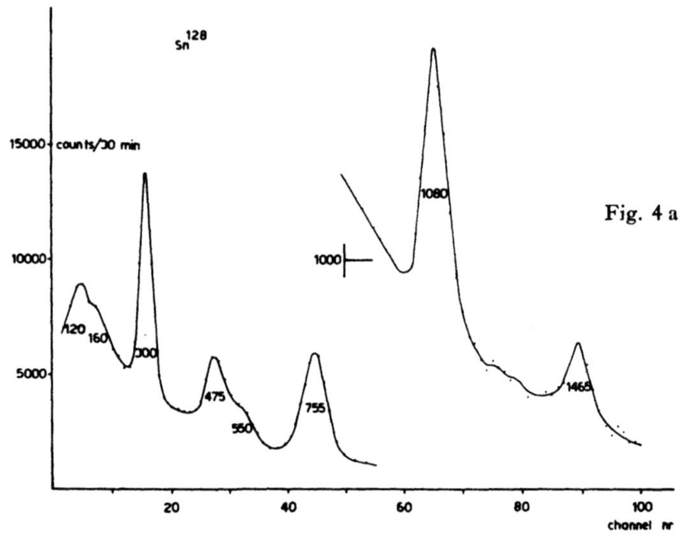

Fig. 4 a

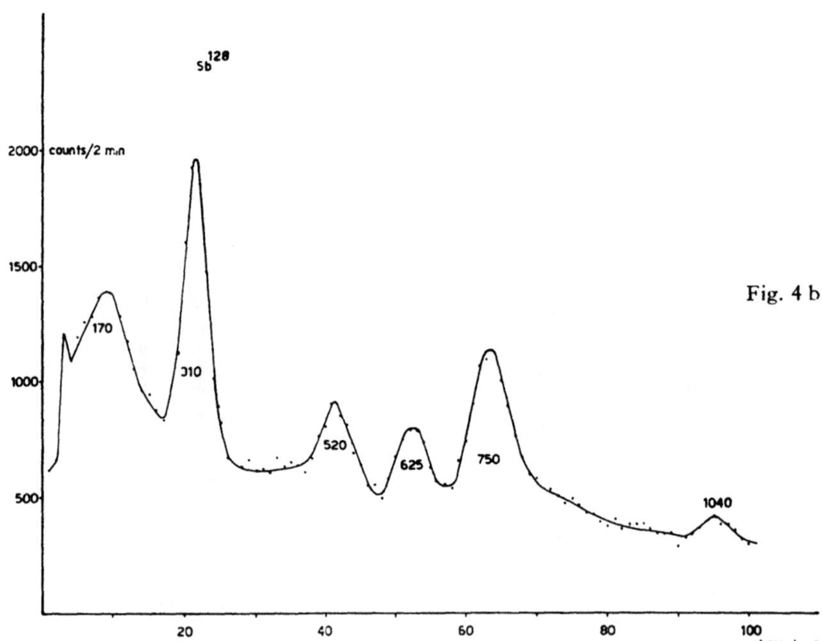

Fig. 4 b

separation for investigations on the other isotopes. The simplest way to perform a check of the separation and determine the exact position of the mass lines is shown in Fig. 3b.

Scattering and ejected electrons give a false distribution, but in general the method gives the information needed concerning position and a rough measure of the peak intensity.

Lastly an example of the need for separated samples in investigating the mass number 128. Sn^{128} decays with a half-life of about one hour to Sb^{128}, which then mainly decays by a shorter half-life to levels in Te^{128}, a stable isotope. Sb^{128}, however, has an isomeric

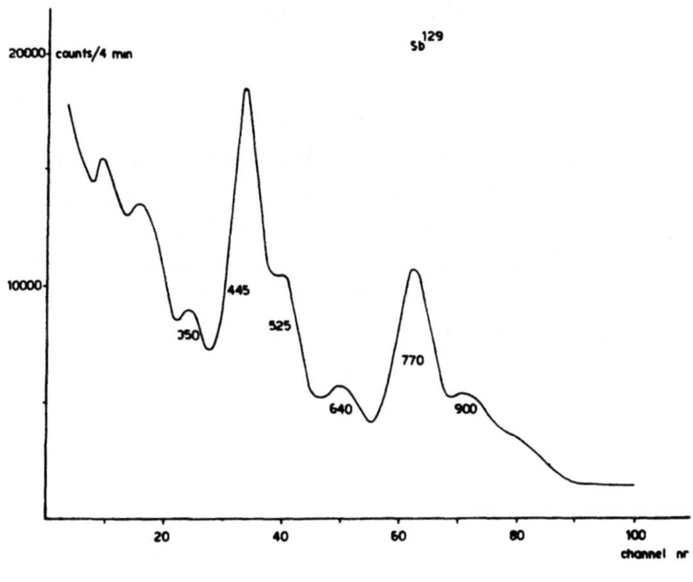

Fig. 4 c

level populated in fission with a nine hour half-life which decays to about the same levels in Te^{128}. To be able to perform measurements on this decay scheme one has to eliminate the much stronger activity of Sb^{129}.

The scintillation spectra, Fig. 4, of these isotopes show the importance of isotope separation for coincidence measurements with a scintillation set up as an example [2].

References

1. Stockendal, R., Ark. Fys. **17**, 553 (1960).
2. Uhler, J., O. Melin, G. Neuman and T. Alväger, to be published in Ark. Fys.

Recent Applications of the Uppsala Mass Separator

By

G. Andersson

The Gustaf Werner Institute for Nuclear Chemistry,
University of Uppsala

With 2 Figures

1. Introduction

The mass separator at the Gustaf Werner Institute [1] was put in operation in 1953. It has been devoted almost exclusively to radioactive work following irradiations in the 190 MeV synchro-cyclotron of the institute. During the first years the separations were relatively rare events, but a reconstruction of the ion source in 1956 [2] increased the efficiency considerably and the separator gradually became a standard tool in our sample preparations.

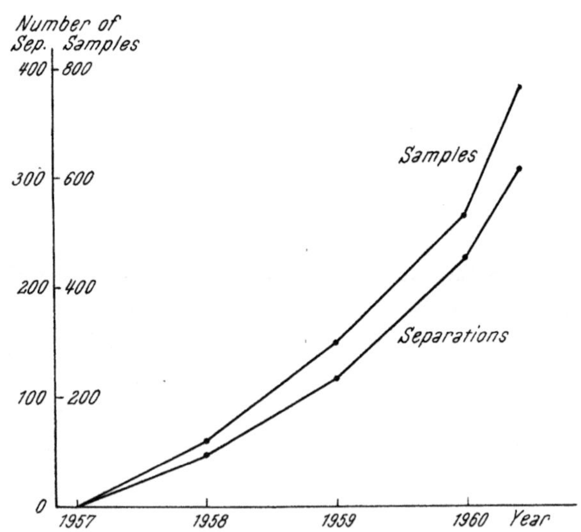

Fig. 1. Operating-statistics of the Uppsala separator for the time covered by this survey. (Cf. the text.)

Fig. 1 gives the cumulated numbers of separations (outer scale) and of utilized samples (inner scale) as functions of time after the reconstruction. The recent steep rise can be correlated with an increase in the separator staff, made about a year ago, from one to two.

The applications fall mainly within the fields of nuclear spectroscopy and reaction yield studies, the elements processed including Zn, Ga, As, Br, Kr, Te, I, Xe, Hg, Tl, and Pb. It is not the intention here to trace the normally quite short life-stories of all the 800 samples prepared since early 1957, but rather to give a general idea of the types of problems concerned and to discuss some special experimental techniques. It may be useful first to say a few words about the separator.

2. Characteristics of the Uppsala separator

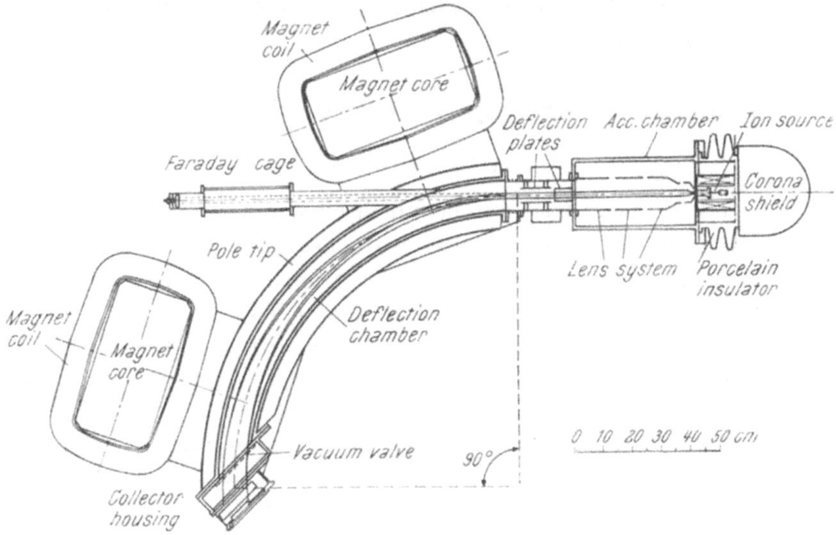

Fig. 2. Design of the Uppsala separator

The main principles of the apparatus are shown in Fig. 2. To quote Dr. Koch's words at the Harwell Conference in 1955, it differed at that time from the Stockholm and Gothenburg separators mainly by having the ions make left turns instead of right turns. By now the differences are much more pronounced, because the other groups have changed over to sector type machines, whereas we have retained the original parallel beam focusing and collect directly at the exit boundary of the magnetic field. The arguments for and against this system cannot be treated in detail here. One positive feature with bearing on some discussions earlier at this Symposium may be worth pointing out, however, namely the absence of stray field effects at the collector side. This means that the focusing conditions are uniform over the whole mass range covered and that the exact positions of the lines are easily calculated and reproducible, which enables us to collect on separate strips placed at predetermined positions. In this way the time of bringing the samples to subsequent measurements is reduced, which is important in much of the work we are doing.

The lines have a height of 20–25 mm and a half-width of 1–2 mm in the focal plane at 45° angle to the beam. As a rule the total ion current lies between 200 and 400 μA. The overall efficiency varies even for the same charge material and has in some cases apparently exceeded the 10% that is considered an average value in actual applications by at least a factor two.

3. Nuclear Spectroscopy of Neutron Deficient Nuclides

3.1. Outlines of the work

Pb and Tl have been our favourite elements for many years. In this mass range the distance between consecutive lines is 3.5 mm in the focal plane, and the contamination from one mass to its neighbours has been estimated as less than 1% at best.

Table 1 shows the extension of our work in the lead region [3–9], including some recently studied Hg and Au isotopes [9, 10]. The

Table 1. *Fragment of a nuclide chart, showing the*

			Pb 194 11 m	Pb 195 17 m	Pb 196 37 m
	Tl 191 < 10 m?	Tl 192 11.4 m | ∼ 11 m?	Tl 193 22.6 m	Tl 194 32.8 m | 33 m	Tl 195 (3.5 s) | 1.16 h
Hg 189 9.5 m	Hg 190 19.8 m	Hg 191 (57 m)			
		Au 190 39m	Au 191 3.2 h		

nuclides in parenthesis are those of which samples have been prepared only for delivery to other laboratories, whereas the parenthesis around a half-life indicates that it has not been measured from a mass separated sample.

A similar project concerning Te and Sb isotopes was started recently. It may be of interest to mention that $TeBr_4$ has proved much better as a charge material in the separator than metallic Te, which was first tried. The results obtained so far [11] establish mass assignments and suggest decay schemes for the following nuclides: 4.5 d Te^{119m}, 16 h Te^{119}, 6.0 d Te^{118}, 61 min Te^{117}, and 2.5 h Te^{116}.

3.2. General characterization of nuclear species

The initial procedures employed in studying earlier unknown nuclides can be summarized as follows:

I. Bombard the element Z with protons of suitable energy in the synchrocyclotron and isolate it chemically, adding a few mg stable carrier, element $Z+1$, formed by (p, xn) reactions, and bring it to a suitable chemical form for mass separation.

II. Mass separate with collection directly into a thin Al strip, mounted on the source holder fitting the two-directional focusing β-spectrometer [12] and placed at a predetermined position in the focal plane of the separator.

extension of the lead region project (Cf. text)

Pb 197		Pb 198		Pb 199		(Pb 200)		(Pb 201)		(Pb 202)	
42 m	< 42 m		2.40 h	12.2 m	1.5 h		21.5 h	(61 s)	9.4 h	3.6 h	($\sim 3 \cdot 10^5$ y)
Tl 196		Tl 197		Tl 198		Tl 199		(Tl 200)		(Tl 201)	
1.41 h	1.8 h	(0.55 s)	2.84 h	1.87 h	5.3 h	(42 ms)	7.42 h		26.1 h	(5 ms)	72 h

III. Scan automatically the conversion electron spectrum repeatedly over a certain energy region, often chosen with regard to energy level systematics.

IV. In a new run, adjust the beta-spectrometer to accept a prominent conversion line and follow its decay down to the background at fixed spectrometer settings.

There are of course deviations from this general scheme. Sometimes the chemical preparation preceding the mass separation can be avoided. Hg isotopes, for instance, are obtained by putting an irradiated Au target directly in the ion source oven together with stable $PbCl_2$, which needs a running temperature that is just suitable for the Hg atoms to leave the target and be ionized with good efficiency. The Pb isotopes are used as mass markers, which means a distance of 17 mass numbers for Pb^{206} from the lightest Hg isotope studied so far.

To investigate Au isotopes, on the other hand, chemical separation is performed after the mass separation of Hg as described above. In this way the Au activities can be obtained free of their Hg parents.

3.3. Beta spectroscopy

The outstanding properties of mass separated sources in beta spectroscopy, especially at low energies, have been pointed out frequently and need not be repeated here. Attention shall instead be drawn to what may be considered a disadvantage of high isotopic purity. In conversion electron spectroscopy, which has played a dominant role in most of our lead region work, good energy determination is often of a decisive importance. This is facilitated by calibration lines, which may be obtained from accurately studied isotopes present in unseparated sources. A successful mass separation removes this possibility.

In our study of Tl^{196} a disagreement as to the energy of an important transition was settled by first collecting Tl^{196} on the beta-spectrometer source backing and then moving the mass spectrum two steps, so as to get Tl^{198}, feeding the very accurately determined 412 keV transition in Hg^{198}, superimposed on the Tl^{196} [7]. Another similar technique consists in collecting the ions on a backing where ThB has been deposited in advance in an activation chamber using a standardized slit position in both cases.

High resolution spectroscopy is always a question of high specific activity, and in this respect the use of mass separated sources in low transmission spectrometers reflects the development of the Scandinavian separators in a rather interesting way. About ten years ago the intensity was in general just sufficient to allow measurements in lens spectrometers at a transmission of about 5% or more and a correspondingly low resolution. Improvements of the focusing properties as well as of the ion source efficiency in the middle of the fifties made mass separation a feasible method of source preparation for two-directional focusing spectrometers. In our Pb and Tl investigations, for instance, where mass separated samples have been used throughout since the beginning of 1957, transmission and resolution have been of the order of 0.2–0.5%. Another step by at least a factor 10 towards lower transmission spectroscopy was taken last year, when we started source preparation for a new semi-circular spectrograph with photographic recording at the neighbouring Physics Department [13]. This type of instrument needs extremely narrow sources, normally deposited on wires, some tenth of a mm thick. The focusing of the mass separator could not reasonably be improved to something like that. Instead a 0.15 mm $\times$ 10 mm slit was put in front of an Al collector strip. In this way about 10% of the total yield of 21 h Pb^{200} was collected, which was quite sufficient to give a line rich spectrum after a 20 h exposure. A comparison of the line shape with that of calibration lines from ThB deposited on a wire revealed the advantage of mass separated sources. The Pb^{200} lines were free of a low energy tail that characterized the calibration lines and could be ascribed to backscattering from the wire.

3.4. Alpha spectroscopy

Besides being essentially carrier free, mass separated sources have the advantage in alpha spectroscopy following high energy reactions, that the beta background is considerably reduced. We have recently, in cooperation with a group from the University of Helsinki, started a search for alpha emitters among the highly neutron deficient isotopes of Pb, Tl, and Hg (cf. Table 1) and also of Te. 5.2 MeV alphas with a branching ratio about $5 \cdot 10^{-7}$ were observed from 20 min Hg^{190}, and some other nuclides gave positive indications, but in most cases the result is merely an upper limit for a possible alpha branch [14]. It should be pointed out that the possibilities of misinterpretations because of impurities in the samples are considerable with the very sensitive ionization chamber method used for the detection of alpha branches weaker than 10^{-6}. Thus it is necessary to study whole series of consecutive isotopes to make sure that a mass assignment is correct.

3.5. Detailed investigations of nuclear levels

A large part of the numerous samples made for the Physics Departments have been used in specialized nuclear studies, such as gamma-gamma and electron-gamma angular correlations and coincidence work of various types. An application of mass separated samples so far unique was the spin measurement of 5.3 h Tl^{198} by means of the atomic beam resonance method [15]. This investigation introduced new aspects on the collection technique, because the samples had to be suitable for evaporation from the oven of the atomic beam apparatus. A promising method of some interest was to collect the Tl^{198} ions in a layer of stable Tl metal vacuum evaporated onto a Pt backing.

4. Reaction yield Studies

The Uppsala separator has right from the beginning been applied to studies of relative yields from spallation and high energy fission reactions [16, 17, 18]. This use of the separator is based on the assumption that there are no mass discriminating effects among isotopes of the same element in the ion source or at the collector. If one aims at high accuracy, this assumption has certainly to be verified or correction factors determined.

The collection efficiency may differ within a series of isotopes if stable ion currents of appreciable strength contaminate some of them, for instance by hydride formation. This was a reason to avoid stable Br as carrier in an investigation of Br and Kr isotopes from high energy fission of uranium [19]. Instead As was chosen as a mass marker but could not be used to sustain the discharge, as it caused frequent sparking in the electrostatic lens system. The problem was solved by running the discharge on AgI placed together with the UO_3 target in the front oven of the ion source [2] and heated to about 500° C, while a small amount of As was put in the rear oven without separate heating.

From the point of view of possible mass discriminating effects, the measurement of isomeric cross section ratios is very attractive, because then only one mass number is concerned. In one such investigation [20] the Br^{80} pair was produced by spallation of yttrium, high energy fission of uranium, and the (p, pn) reaction on bromine, and the yield ratio studied as a function of bombarding energy. The results differ in a significant way: in the last-mentioned case the ratio is energy-independent, whereas spallation and fission show decreasing and increasing ratios, respectively, when the energy increases from 70 to 160 MeV. Spallation yield ratios of isomeric pairs have been further studied in a separate investigation [21], also largely based on mass separated samples.

References

1. ANDERSSON, G., Ark. Fys. **12**, 331 (1957).
2. BURGMAN, J. O., and G. ANDERSSON, Nucl. Instr. **3**, 33 (1958).
3. ANDERSSON, G., E. ARBMAN, and B. JUNG, Ark. Fys. **11**, 297 (1957).
4. ANDERSSON, G., P. A. TOVE, B. JUNG, and I. B. SVENSSON, Nuclear Physics **3**, 493 (1957).
5. ANDERSSON, G., J. O. BURGMAN, and B. JUNG, Phil. Mag. **3**, 105 (1958).
6. JUNG, B., Nuclear Physics **10**, 440 (1959).
7. JUNG, B., and G. ANDERSSON, Nuclear Physics **15**, 108 (1960).
8. JUNG, B., and J. SVEDBERG, Nuclear Physics **20**, 630 (1960).
9. ANDERSSON, G., I. B. HÄLLER, and R. RINGH, J. Inorg. Nucl. Chem. in press.
10. ANDERSSON, G., and R. RINGH, Ark. Fys. **18**, 385 (1960).
11. FINK, R. W., G. ANDERSSON, and J. KANTELE, Ark. Fys. in press.
12. ARBMAN, E., and N. SVARTHOLM, Ark. Fys. **10**, 1 (1956).
13. KARLSSON, E., and K. SIEGBAHN, Nucl. Instr. **7**, 113 (1960).
14. KARRAS, M., G. ANDERSSON, and M. NURMIA, to be published.
15. LINDGREN, I., C. M. JOHANSSON, and S. AXENSTEN, Phys. Rev. Letters **1**, 473 (1958).
16. ANDERSSON, G., Phil. Mag. **45**, 621 (1954).
17. RUDSTAM, G., Thesis, University of Uppsala (1957).
18. AAGAARD, P., G. ANDERSSON, J. O. BURGMAN, and A. C. PAPPAS, J. Inorg. Nucl. Chem. **5**, 105 (1957).
19. HAGEBÖ, E., Cand. real. thesis, University of Oslo (1959) (unpublished).
20. HÄLLER, I. B., and G. ANDERSSON, J. Inorg. Nucl. Chem. in press.
21. HÄLLER, I. B., and G. RUDSTAM, J. Inorg. Nucl. Chem. in press.

Applications in the Field of Nuclear Physics of the Orsay Double Deflexion Isotope Separator

By

R. Bernas

Université de Paris, Laboratoire de Physique Nucléaire, Orsay (S. et O.)

With 4 Figures

Summary

We shall give here a brief review of experiments done with the help of the first or both stages of the Orsay separator during its first year of operation.

In this short summary we shall not enter into the details of the nuclear physics problems studied but rather give a review on the possibilities of an instrument of this type.

I. Relative Cross-Section Measurements for the Production of Radioactive Mercury Isotopes from Gold [1]

Several authors have examined (p, xn) spallation reactions but very few experimental results have been given of cross-sections for more than two isotopes of an element: one may refer to work on bismuth [2], caesium [3] and ytterbium [4].

The radiochemical studies used in these experiments make it difficult to identify isotopes with similar half-lives. The problem is greatly facilitated by the electromagnetic separation of the radioactive isotopes.

Andersson and col. [5] have used this method in the past for cross-section measurements and for nuclear spectroscopic studies of isotopes produced by spallation reactions.

The gold targets used in our experiments were usually bombarded for 10 min by a 1 to 2 μA beam of 155 MeV protons from the Orsay synchrocyclotron and then separated.

A. Separation with double deflection

In preliminary experiments we studied separately a few mercury isotopes and we determined an absolute cross-section value which was later used as a standard for relative measurements.

The mass identification procedure described previously [6] was confirmed by the study of Hg^{191}, which is well known.

The isotopes studied decayed mainly by a K capture process and they were detected by the use of NaI (Tl) scintillators and a 256 channel analyser. The results of measurements are shown below:

Table 1

Mass	Half-lives and main γ lines (keV)
191	$\mathrm{Hg} \xrightarrow[(260)]{55\ \mathrm{min}} \mathrm{Au} \xrightarrow[(140,300,600)]{3.4\ \mathrm{h}} \mathrm{Pt} \xrightarrow[(360,530)]{3\ \mathrm{d}}$ Ir stable
190	$\mathrm{Hg} \xrightarrow[(140,220)]{21\ \mathrm{min}} \mathrm{Au} \xrightarrow[(290,600)]{45\ \mathrm{min}}$ Pt stable
189	$\mathrm{Hg} \xrightarrow[(165,2403,20,500)]{9\ \mathrm{min}} \mathrm{Au} \xrightarrow{30\ \mathrm{min}} \mathrm{Pt} \xrightarrow[(150)]{11.5\ \mathrm{d}} \mathrm{Ir} \xrightarrow[(250)]{12.5\ \mathrm{d}}$ Os stable
188	$\mathrm{Hg} \xrightarrow[(140)]{3.7\ \mathrm{min}} \mathrm{Au} \xrightarrow[(250,230,630)]{8\ \mathrm{min}} \mathrm{Pt} \xrightarrow[(190)]{10\ \mathrm{d}} \mathrm{Ir} \xrightarrow[(150,480,630)]{40\ \mathrm{h}}$ Os stable

B. Simultaneous collection of isotopes at the first stage

The measurement of relative cross sections of (p, xn) reactions was determined after three simultaneous collections of 10 mercury isotopes at the collector of the first stage.

The results are shown on Fig. 1 and details as well as interpretation are given in the paper quoted above [1].

The simultaneous collection of isotopes at the collector of the second stage is being carried out at present and should cover a total mass-range of 5% at a time.

II. Studies on Isotopes with Short Half-Lives

The possibility of replacing the normal collector at the second stage by a scintillator and a photomultiplier has been described previously [6].

We have made use of this property in the study of the mercury isotopes of mass 187, 186, and 185. The experimental set up is shown elsewhere[1].

The pulses, after amplification were sent simultaneously to a multi-channel selector for the γ-ray analysis and to a single-channel one, for the study of the K capture decay.

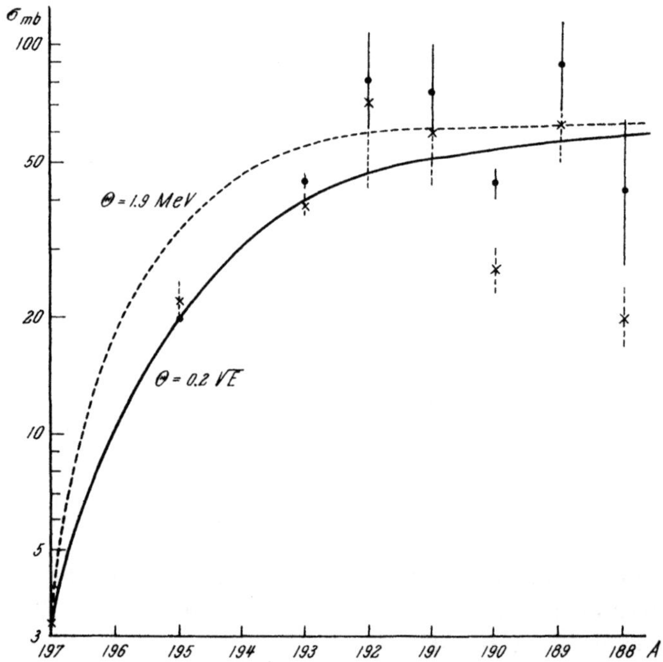

Fig. 1. Cross-section for production of Hg isotopes from gold by 155 MeV protons

The gold targets were bombarded by the external beam of the 155 MeV Orsay synchro-cyclotron and isotopic separation started less than 3 minutes after the end of the irradiation. The time required for the deposition of the mercury ions was only 30 seconds (see ascending first part of decay curve Fig. 2).

[1] See Fig. 4 of "An Ion Source for Short-Lived Radioactive Isotopes And Preliminary Results on a Duoplasmatron Type Ion Source" (page 184).

The analysis of the γ-spectra and of the decay curves led to the results summarized in Table 2 and detailed results are given in reference [7].

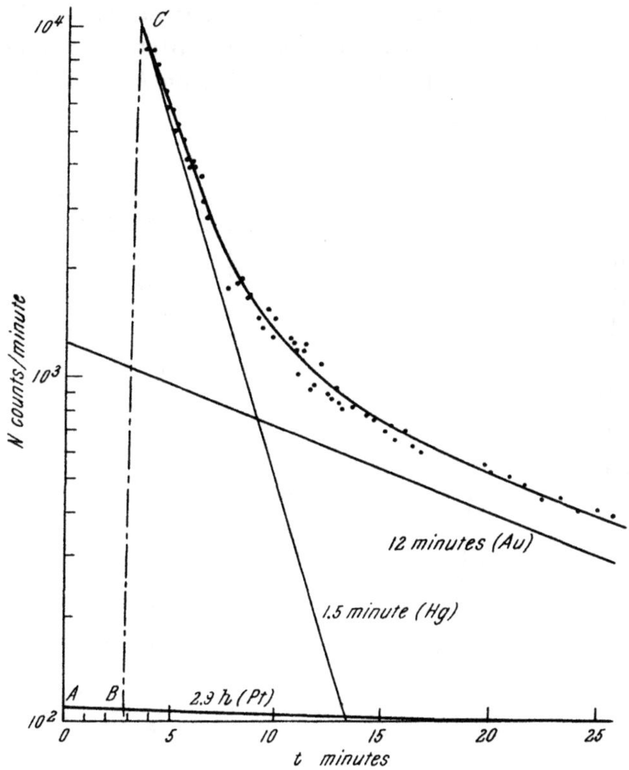

Fig. 2. Build up and decay curve for Hg^{186} at the collector of the separator
A. End of cyclotron run B, C. Accumulation of Hg^{186}

Table 2

Masses	Half-lives and main γ-lines					
	Hg →		Au →		Pt	
	T	γ(keV)	T	γ(keV)	T	γ(keV)
188	3.7 m	140	8 m	250, 330, 630	10 d	190
187	3 m	175, 255, 400	8 m	—	2.2 h	
186	1.5 m	125, 270, 350, 440	12 m	160, 220, 300, 400	2.9 h	
185	50 s		7 m		1.2 h	

III. Preparation of Sources for β-Spectroscopy

The conversion electron spectra from the 28.4 keV transition in Au^{190} and from the 31.4 keV transition in Au^{192} are shown on Fig. 3 as an example of the well-known quality of electromagnetically separated sources prepared by double magnetic deflexion. Due to the relatively short half-lives of theses nuclides repeated separations were performed every two hours. Detailed studies on Au^{190} and Au^{192} will be published shortly [8].

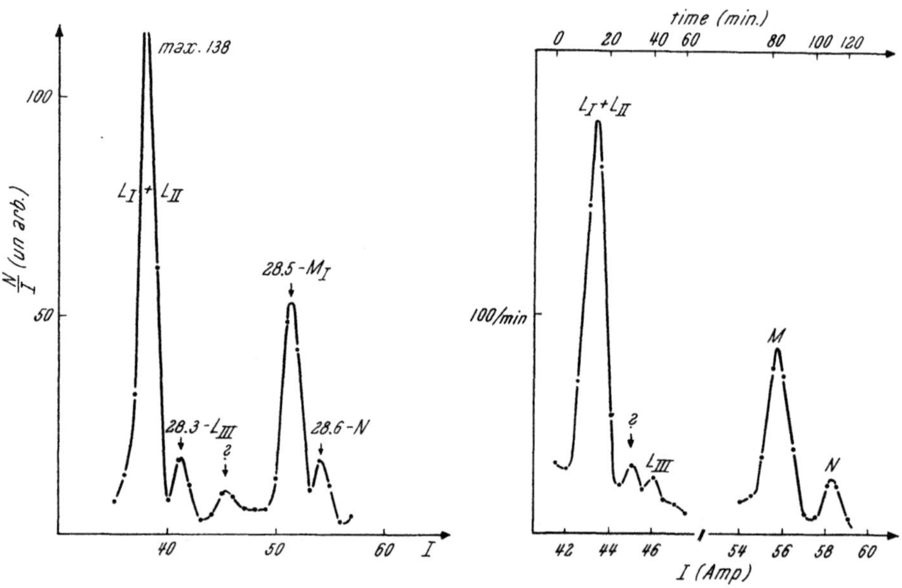

a b

Fig. 3. β-spectrometer records of two transitions in Au^{190} and Au^{192} obtained from separated samples

a 31.4 keV Au^{192} b 28.4 keV Au^{190}

Sources of $Hg^{191, 192, 193, 195}$ for the study of low energy electrons (Auger and conversion) in a semi-circular magnetic spectrograph were prepared after double deflexion. The requirements of high intensity and a very thin deposit were met. Fig. 4 shows the K, L, M Auger electron ($0 < E < 80$ keV) as well as some internal conversion lines of $^{195}Au + ^{195}Pt$(9). This spectrum was obtained with the preacceleration method ($V_{acc} = 10$ kV).

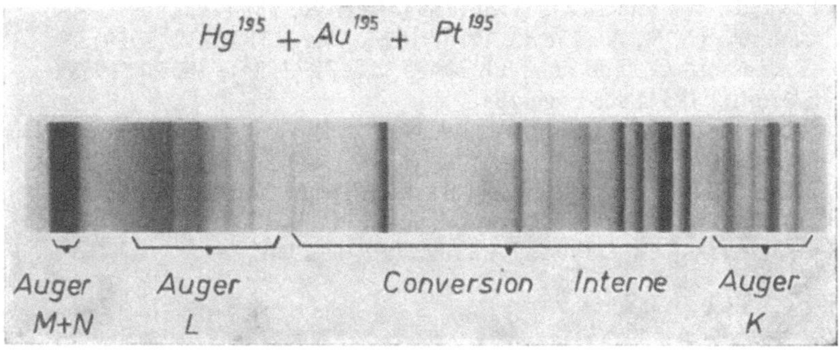

Fig. 4. *K*, *L*, *M* Auger electron and internal conversion lines of $^{195}Au + ^{195}Pt$ from a separated ^{195}Hg target

IV. Use of High Purity Tracers in Isotope Dilution Work

One of the most important applications of the double magnetic deflexion separator is the production of very high purity isotopes for use as tracers in the isotope dilution method. A programme of work in this field is being started in our laboratory and two experiments are in progress:

a) cross-section measurement for the production of Li by high energy proton bombardment of oxygen,

b) cross-section the production of Yb isotopes by spallation reactions on tantalum.

As a preliminary test for the first measurement we determined [10] the cross section of the reaction

$$V^{51}\ (n, \gamma)\ V^{52} \xrightarrow[3.7\ \text{min}]{\beta^-} Cr^{52}\ (\text{stable})$$

and made a comparison with the accepted value obtained by activity measurements. The use of highly enriched Cr^{53} as a tracer made this experiment possible.

Yb^{174} and Yb^{171} of high purity are also required for the last experiment.

In all isotope dilution experiments the availability of very pure tracers will increase the sensitivity of the method and only small amounts will be necessary so that these can be obtained readily with a laboratory separator such as this one.

References

1. Poffé, N., and al., J. Phys. Radium **21**, 343 (1960).
2. Hunter, E. T., and J. M. Miller, Phys. Rev. **115**, 1053 (1959).
3. Fink, R. W., and E. O. Wiig, Phys. Rev. **96**, 185 (1954).
4. Caretto Jr., A. A., and O. Wiig, Phys. Rev. **115**, 1238 (1959).
5. Andersson, G., and al., Phil. Mag. **45**, 621 (1954); **46**, 70 (1955).
6. Sarrouy, J. L., see page 184.
7. Albouy, G., M. Gusakow, and N. Poffé, C. R. Ac. Sciences (Paris, Oct. 1960).
8. Mme P. Kilcher and J. Jabstrevsky, Private Communication.
9. Dionisio T. S., Private communication.
10. Gradsztajn, E., J. Phys. Radium **21**, 761 (1960).

Preparation of Radioactive Sources for β- and γ-Ray Spectroscopy by Electromagnetic Isotope Separation

By

O. B. Nielsen and **O. Skilbreid**

Institute for Theoretical Physics, Copenhagen

Summary

Some of the problems related to the preparation of sources for nuclear spectroscopy are discussed. A number of experiments are described, which could not have been performed without the use of electromagnetic isotope separators.

In the past, the isolation of radioactive material and the subsequent preparation of suitable sources for investigations of the β- and γ-radiations emitted by the active isotopes have generally been carried out by radiochemical means. The chemical methods, however, have an essential limitation due to the fact that radioactive isotopes are usually not produced isolated from other radioactive or stable isotopes of the same element. Thus a larger number of isotopes are formed simultaneously not only in fission, but also in the reactions initiated by high energy charged particles in accelerators, in which case a varying number of neutrons evaporate from the nucleus. In a reactor, more isotopes can be produced when the target element is not monoisotopic, and the active sample furthermore consists mainly of the stable target material.

Although, frequently, samples sufficiently good for spectroscopy can be obtained of the long-lived members of the mixture after the decay of the short-lived, isotope separation represents in many cases the only possibility of providing pure sources, apart from being the only really safe method of assigning the radiations observed to definite mass numbers. Fortunately enough the technique of electromagnetic isotope separation has made such progress that separated radioactive sources are available of almost every element. As a consequence, it has been possible to study many activities which would otherwise have been obscured by the presence of other radiations. In addition it has turned out that sources prepared in an electromagnetic isotope separator are of such a high quality that this technique is also preferable in cases where a real isotope separation is not essential.

The high quality of the separated sources is especially important for investigations in β-ray spectrometers. For measurements with high resolution a source thickness corresponding to only a few $\mu g/cm^2$ is required. Although an average thickness of this order is often obtained when a carrier-free activity is deposited by chemical means, the effective thickness is normally much higher due to the formation of crystals. In an isotope separator, the activity is uniformly distributed on a small focal spot, the effective thickness being just a few $\mu g/cm^2$, determined by the penetration of the ions in the target surface. The method is thus applicable, e. g., to the preparation of high quality β-ray sources from reactor-produced samples of low specific activity. Intense β-ray sources even of the noble gases such as xenon, krypton, and radon have been made, since the atoms of these elements can be trapped in the target foil, when this is hit by the fast ions. The electromagnetic separation technique thus provides the only method suited for preparing β-ray sources of the noble gases.

Until a few years ago, the efficiency of radioactive isotope separations was in general quite low, but recent advances in ion-source techniques, e. g., the chlorine micromethod and the application of two ovens, have improved the situation remarkably, efficiencies of 5–10% being quite normal, and 20–40% have been reported in some cases. This question is obviously the one most deserving of serious consideration at the present stage of development, also because its high efficiency reduces health hazards.

Another point of extreme importance for the work with radioactive isotopes is the time needed for preparing the charge for the ion source, and the separation itself. The new ion source techniques are also in this respect real improvements, since sources can be ready for measurements less than 1 hour after the arrival of the activity. The method is now useful for activities with half-lives in the minutes region.

Obviously, the radioactive isotopes represent minute amounts by weight, and the separations therefore can take place in smaller separators of the laboratory type, which can operate with an isotopic enrichment well above 100. The purity of the separated samples is thus almost ideal, actually 99% or better. The small amounts of material needed for loading the ion source also make it possible in many cases to use material for activation which was already enriched in a calutron. The smaller types of separators thus rather supplement than supplant the calutrons.

In studies of radioactive decay schemes performed in Copenhagen, sources of the following elements have been prepared: In, Sb, Nd, Sm, Gd, Tb, Ho, Tm, Hf, Re, Hg, Em, Th, Pa, U, and Pu. Some typical examples of problems which could not have been solved without the aid of isotope separators will be briefly described below.

Pu^{242}. We had at our disposal a sample consisting mainly of Pu^{242}, but with an impurity of Pu^{241}, which was small by weight. Due to the shorter half-life of Pu^{241}, this isotope contributed most of the radiation from the sample, but we could not wait until the short-lived component had died, since the half-lives are 13 years and $3.8 \cdot 10^5$ years. An isotope separation with 5% efficiency produced a pure Pu^{242} source sufficiently strong for the measurement of the lifetime of the first excited level in the daughter U^{238}.

Th^{232}. Natural Thorium exists in equilibrium with the descendant Th^{228}, which contributes strongly to the α- and γ-radiation from normal Th samples, and a purification cannot be performed with chemical means. A 1 mg sample of Th^{232} depleted in Th^{228} was made in a 20 h run in the isotope separator. This sample was strong enough for the measurement of the lifetime of the first excited level in Ra^{228}.

RaB and RaC. Very strong and thin sources are easily available of these classical activities, but good β-ray spectra were never obtained due to the short half-lives, which are below $\frac{1}{2}$ hour for both isotopes. The difficulties were overcome by preparing strong sources of the grandmother Em^{222} in the isotope separator by letting the ion beam impinge on a thin foil. Em^{222} decays by two α-decays, which do not interfere with the β-ray measurements, into RaB, which is thus formed continuously on the source, the effective half-life now being that of Em^{222} or 4 days. This example is thus a case where no isotope separation was needed, but only an isotope separator.

Gd^{159}. This activity is formed by neutron capture in Gd^{158}, but only with a low cross section, whereas neighbouring rare earth elements, which are always present as small impurities in Gd, are activated with a very high cross section. Isotope separation procured sources which were pure as well as of high specific activity.

Tb^{151} and Tb^{152} are produced simultaneously by the Gd(p,xn) reaction. The half-lives are almost identical, 18.5 hours, and the radiations from the two isotopes could not be assigned to a definite mass number before isotope separation.

The isotope separator is thus essential for solving a large variety of problems in nuclear spectroscopy. There are, however, some main difficulties in introducing electromagnetic isotope separation as a standard method in every laboratory studying radioactivity. The equipment must be of a high technical standard, and much experience is needed by the operators. The ion source technique is still at such a stage of development that a certain effort must be made to develop a suitable method for each new element. In this respect, the physical and chemical similarity of the rare earths, so troublesome for the chemist, have been a great advantage, since it was possible to apply the same method to all the elements.

Author Index

Subject Index

Druck: R. Spies & Co., Wien V.

GPSR Compliance

The European Union's (EU) General Product Safety Regulation (GPSR) is a set of rules that requires consumer products to be safe and our obligations to ensure this.

If you have any concerns about our products, you can contact us on ProductSafety@springernature.com

In case Publisher is established outside the EU, the EU authorized representative is:

Springer Nature Customer Service Center GmbH
Europaplatz 3
69115 Heidelberg, Germany

Batch number: 10577878

Printed by Printforce, the Netherlands